GCSE IN A WEEK

CHEMISTRY

Emma Poole

Revision Planner

Page	Day	Time ( )	Title	Exam Board	Date	Time	Completed
Day 1							
4	1	15 mins	How Science Works	AEO			
6	1	15 mins	Atomic Structure	AEO			
8	1	15 mins	The Periodic Table	AEO			
10	1	15 mins	Electronic Structure	AEO			
12	1	15 mins	Ionic Bonding	AEO			
14	1	15 mins	Covalent Bonding	AEO			
16	1	15 mins	Alkali Metals	AEO			
Day 2							
18	2	15 mins	Noble Gases and Halogens	AEO			
20	2	15 mins	Calculations 1	AEO			
22	2	15 mins	Calculations 2	AEO			
24	2	15 mins	Calculations 3	AEO			
26	2	15 mins	Haber Process	AEO			
28	2	15 mins	The Manufacture of Sulfuric Acid	AEO			
Day 3							
30	3	15 mins	Rates of Reaction	AEO			
32	3	15 mins	Energy	AEO			
34	3	15 mins	Calorimetry	AEO			
36	3	15 mins	Energy Profile Diagrams	AEO			
38	3	15 mins	Aluminium	AEO			
40	3	15 mins	Sodium Chloride	AEO			
42	3	15 mins	Copper	AEO			
Day 4							
44	4	15 mins	Acids and Bases	AEO			
46	4	15 mins	Making Salts	AEO			
48	4	15 mins	Metals	AEO			
50	4	15 mins	Useful Metals	AEO			
52	4	15 mins	Iron and Steel	AEO			
54	4	15 mins	Water	AEO			

Page	Day	Time (mins)	Title	Exam Board	Date	Time	Completed
Day 5							
56	5	15 mins	Hard and Soft Water	AEO			
58	5	15 mins	Detergents	AEO			
60	5	15 mins	Special Materials	AEO			
62	5	15 mins	New Materials	AEO			
64	5	15 mins	Vegetable Oils	AEO			
66	5	15 mins	Food Additives	AEO			
68	5	15 mins	Crude Oil	AEO			
Day 6							
70	6	15 mins	Fuels	AEO			
72	6	15 mins	Air and Air Pollution	AEO			
74	6	15 mins	Pollution	AEO			
76	6	15 mins	Alkanes and Alkenes	AEO			
78	6	15 mins	Alcohols, Acids and Esters	AEO			
Day 7							
80	7	15 mins	Polymers	AEO			
82	7	15 mins	Limestone	AEO			
84	7	15 mins	Structure of the Earth	AEO			
86	7	15 mins	Cosmetics	AEO			
88	7	15 mins	Chemicals	AEO			
90	Answers						

How Science Works

Sometimes our opinions are based on our own prejudices; what we personally like or dislike.

Ideas

At other times, our opinions can be based on scientific evidence. These opinions are based on reliable and valid evidence that can be used to back up our opinion.

Variables

- An **independent** variable is the variable that we choose to change to see what happens.
- A **dependent** variable is the variable that we measure.
- A **continuous** variable, e.g. time or mass, can have any numerical value.
- An **ordered** variable, e.g. small, medium or large, can be listed in order.
- A **discrete** variable can have any value which is a whole number, e.g. 1, 2.
- A **categoric** variable is a variable that can be labelled, e.g. red, blue.
- We use **line graphs** to present data where the independent variable and the dependent variable are both continuous. A line of best fit can be used to show the relationship between variables.
- **Bar graphs** are used to present data when the independent variable is categoric and the dependent variable is continuous.

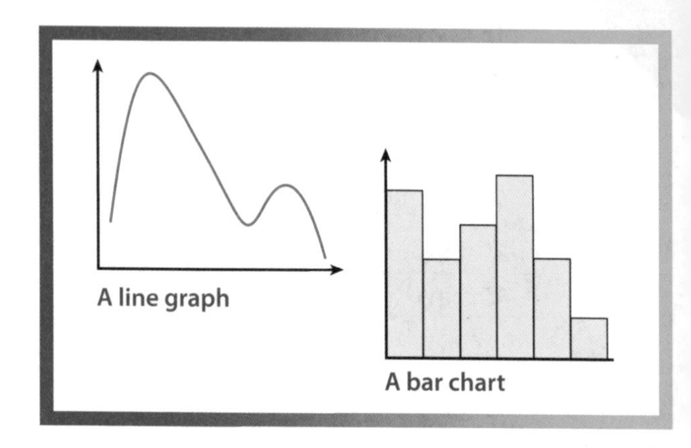

Evidence

Evidence should be:

- **reliable** (if you do it again you get the same result)
- **accurate** (close to the true value).

Scientists often try to find links between variables.

Links can be:

- causal – a change in one variable produces a change in the other variable
- a chance occurrence
- due to an association, where both of the observed variables are linked by a third variable.

We can use our existing models and ideas to suggest why something happens. This is called a **hypothesis**. We can use this hypothesis to make a **prediction** that can be tested. When the data is collected, if it does not back up our original models and ideas, we need to check that the data is valid, and if it is we need to go back and change our original models and ideas.

Science in Society

Sometimes scientists investigate subjects that have social consequences, e.g. food safety. When this happens, decisions may be based on a combination of the evidence and other factors, such as bias or political considerations.

Although science is helping us to understand more about our world there are still some questions that we cannot answer, such as: is there life on other planets? Some questions are for everyone in society to answer, not just scientists, such as: should we clone people?

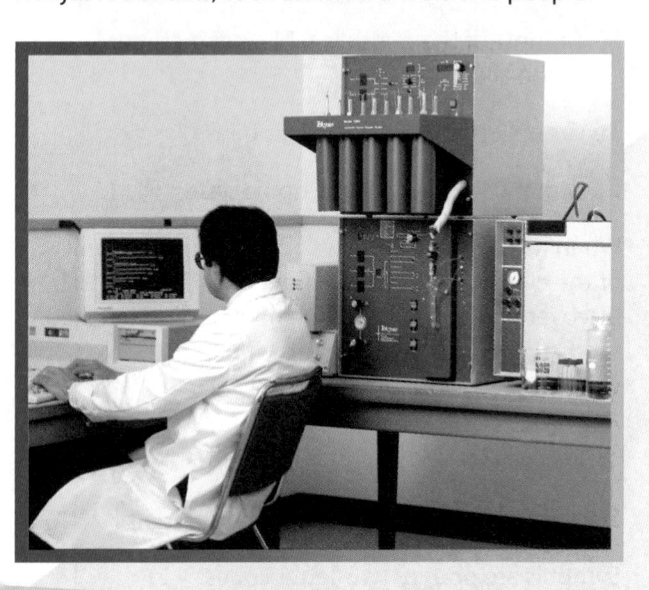

PROGRESS CHECK

1. What is an independent variable?
2. What is a dependent variable?
3. What is an ordered variable?
4. What is a discrete variable?
5. What does accurate mean?

EXAM QUESTION

A student carries out an experiment to find out how the force applied to a spring affects the length of the spring.

a. What is the independent variable?

b. What is the dependent variable?

c. Suggest a variable that must be controlled to make it a fair test.

Atomic Structure

Elements are made of only one type of **atom**. Atoms consist of a small, central **nucleus** (which contains protons and neutrons) surrounded by shells of electrons.

Atoms

All atoms of the same element have the same number of protons.

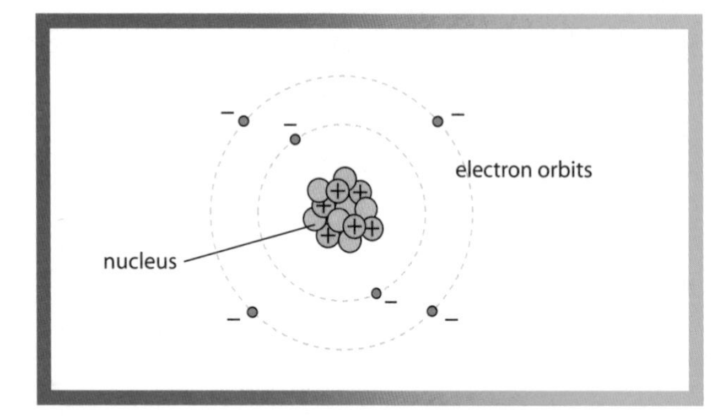

Protons have a mass of 1 and a charge of 1^+. Electrons have a negligible mass and a charge of 1^-. Neutrons have a mass of 1 and no charge.

Bonding

Atoms can join together by:

- sharing electrons (this is called a covalent bond)
- transferring electrons (this forms ions, the attraction between oppositely charged ions is called an ionic bond).

The Periodic Table

- There are about 100 different types of element.
- These elements are often displayed in the periodic table.
- The horizontal rows are called **periods**.
- The vertical columns are called **groups**.
- Group 1 of the periodic table is sometimes known as the alkali metals.
- Group 7 is known as the halogens.
- Group 0 is known as the noble gases.

When the periodic table was first designed, many of the elements which we know today had yet to be discovered. Gaps were left in the table and detailed predictions were made about what the new elements would be like.

Symbols

Elements can be represented using symbols.

Symbols are one- or two-letter codes:

- O is used for oxygen
- C is used for carbon
- Ca is used for calcium
- K is used for potassium
- Na is used for sodium
- Pb is used for lead
- Fe is used for iron.

Formula

The chemical formula tells us the number and type of atoms present.

- An oxygen molecule has the formula O_2. This tells us that an oxygen molecule consists of two oxygen atoms.
- A nitrogen molecule has the formula N_2. This tells us that a nitrogen molecule consists of two nitrogen atoms.
- Carbon dioxide has the formula CO_2. This tells us that a carbon dioxide molecule consists of one carbon atom and two oxygen atoms.
- Water has the formula H_2O. This tells us that a water molecule consists of two hydrogen atoms and one oxygen atom.
- Copper sulfate has the formula $CuSO_4$. This tells us that the ratio of the atoms is one copper atom: one sulfur atom:four oxygen atoms.
- Calcium hydroxide has the formula $Ca(OH)_2$. This tells us that the ratio of the atoms is one calcium atom : two oxygen atoms : two hydrogen atoms.

Progress Check

1. What is special about atoms of the same element?
2. What is the small, central part of an atom called?
3. What are found in shells around the central part of an atom?
4. How are the names and symbols of elements often displayed?
5. Roughly how many elements are there?

Equations

We can use word equations to sum up what happens during a chemical reaction.

For example, when carbon is burned in plenty of oxygen, carbon dioxide is produced. This can be written as:

carbon + oxygen → carbon dioxide

This reaction can also be summed up using the symbol equation:

$$C + O_2 \rightarrow CO_2$$

When hydrogen is burnt in oxygen water vapour is produced. This can be written as

hydrogen + oxygen → water vapour

This can also be written as

$$H_2 + \frac{1}{2}O_2 \rightarrow H_2O$$

During a chemical reaction, atoms are not created or destroyed, they are simply rearranged. This means that there must be equal numbers of each type of atom on both sides of the equation.

Exam Question

Calcium carbonate, $CaCO_3$, reacts with hydrochloric acid to produce the salt calcium chloride, $CaCl_2$, water, H_2O and carbon dioxide, CO_2.

a. Write a word equation for the reaction.

b. Write down the type and number of atoms present in one molecule of carbon dioxide.

The Periodic Table

In the modern periodic table there are about 100 different types of element, the horizontal rows are called **periods**, the vertical columns are called **groups**.

Elements with similar properties occur periodically so it is called the periodic table.

As the elements were discovered, scientists tried to place them into a logical order. John Newlands and Dmitri Mendeleev tried to place the elements in order of increasing atomic weight.

Mendeleev realised that, although this was a good basis, if he stuck to this order too rigidly it did not work. When this happened, he swapped the order of the elements or left gaps. When gaps were left he made detailed predictions about what the new elements would be like.

When the elements were eventually discovered and their properties compared with the predictions Mendeleev had made, they matched very well and it proved how powerful the periodic table was.

At the start of the twentieth century, scientists developed the technology required to discover protons, neutrons and electrons. When they looked back at Mendeleev's table, they realised that by occasionally swapping the order of the elements, what he had actually done was to place the elements in perfect order of increasing atomic number (number of protons).

In the modern periodic table, we say that the elements are arranged in order of increasing atomic number. Elements in the same group (column) have the same number of electrons in their outer shells.

Mass Number and Atomic Number

Two numbers are often written next to the symbols, these are the mass number and the atomic number.

The mass number tells us the number of protons added to the number of neutrons in the nucleus of one atom.

The atomic number, which is sometimes called the proton number, tells us the number of protons in the nucleus of an atom. In a neutral atom this also tells us the number of electrons.

An atom of sodium has a mass number of 23 and an atomic number of 11.

This means that this atom of sodium has:

- 11 protons (atomic number)
- 11 electrons (atomic number)
- 12 neutrons (mass number – atomic number).

An atom of oxygen has a mass number of 16 and an atomic number of 8. This means that this atom of oxygen has:

- 8 protons (atomic number)
- 8 electrons (atomic number)
- 8 neutrons (mass number – atomic number).

Group

Period	1	2											3	4	5	6	7	0
1	H Hydrogen																	He Helium
2	Li Lithium	Be Beryllium											B Boron	C Carbon	N Nitrogen	O Oxygen	F Fluorine	Ne Neon
3	Na Sodium	Mg Magnesium											Al Aluminium	Si Silicon	P Phosphorous	S Sulfur	Cl Chlorine	Ar Argon
4	K Potassium	Ca Calcium	Sc Scandium	Ti Titanium	V Vanadium	Cr Chromium	Mn Manganese	Fe Iron	Co Cobalt	Ni Nickel	Cu Copper	Zn Zinc	Ga Gallium	Ge Germanium	As Arsenic	Se Selenium	Br Bromine	Kr Krypton
5	Rb Rubidium	Sr Strontium	Y Yttrium	Zr Zirconium	Nb Niobium	Mo Molybdenum	Tc Technetium	Ru Ruthenium	Rh Rhodium	Pd Palladium	Ag Silver	Cd Cadmium	In Indium	Sn Tin	Sb Antimony	Te Tellurium	I Iodine	Xe Xenon
6	Cs Caesium	Ba Barium	La Lanthanum	Hf Hafnium	Ta Tantalum	W Tungsten	Re Rhenium	Os Osmium	Ir Iridium	Pt Platinum	Au Gold	Hg Mercury	Tl Thallium	Pb Lead	Bi Bismuth	Po Polonium	At Astatine	Rn Radon
7	Fr Francium	Ra Radium	Ac Actinium															

Isotopes

Isotopes are different forms of the same element. This means that they have the same number of protons but a different number of neutrons.

1. What are the vertical columns in the periodic table called?
2. What are the horizontal rows in the periodic table called?
3. What does the mass number tell us?
4. What is the same about isotopes of an element?
5. What is different about isotopes of an element?

EXAM QUESTION

An atom of carbon has a mass number of 13 and an atomic number of 6.

How many protons, electrons and neutrons does this atom of carbon have?

Electronic Structure

The electronic structure of an atom determines its characteristics and chemical nature.

Atoms

Atoms consist of a small, central **nucleus** (which contains protons and neutrons) surrounded by shells of **electrons**.

Electrons

Electrons fill up the shell closest to the nucleus (called the first shell) first. When this is full, they fill up the next shell.

In our model, there is room for up to two electrons in the first shell and up to eight electrons in the other shells.

Examples

A lithium atom has three electrons – two in the first shell and one in the second shell, giving it an electronic structure of 2,1.

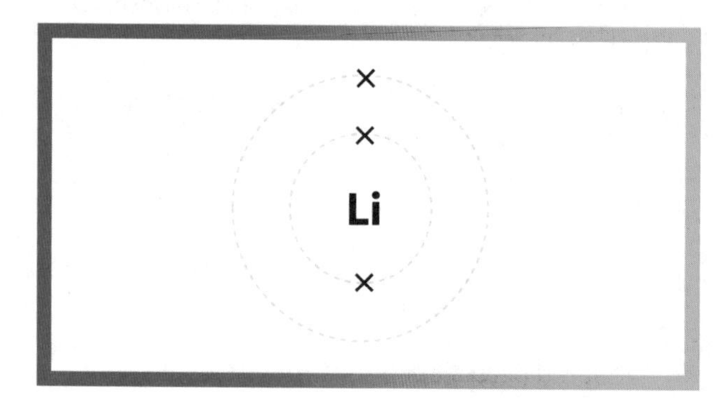

A sodium atom has eleven electrons – two in the first shell, eight in the second shell and one in the third shell, giving it an electronic structure of 2,8,1.

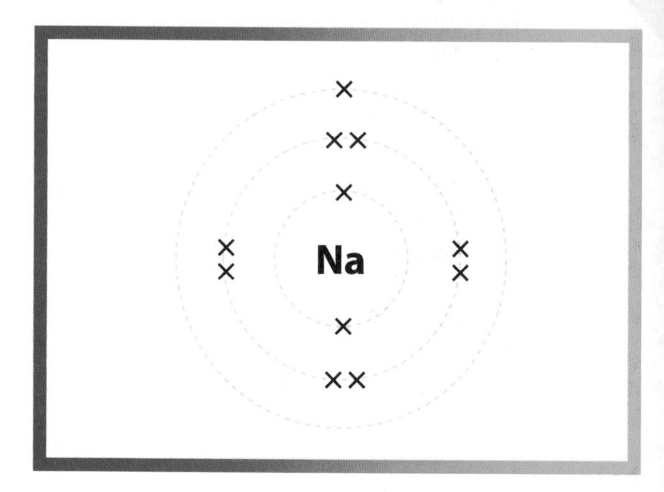

The number of electrons in the outer shell of an atom reveals the group of the periodic table that the element belongs to.

Lithium and sodium both have just one electron in their outer shell, so both belong to Group 1 of the periodic table.

A carbon atom has six electrons – two in the first shell and four in the second shell, giving it an electronic structure of 2,4.

Carbon belongs to Group 4 of the periodic table.

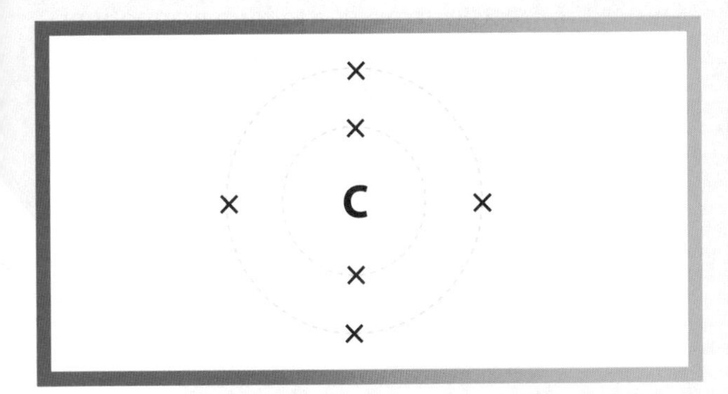

An oxygen atom has eight electrons – two in the first shell and six in the second shell, giving it an electronic structure of 2,6.

Oxygen belongs to Group 6 of the periodic table.

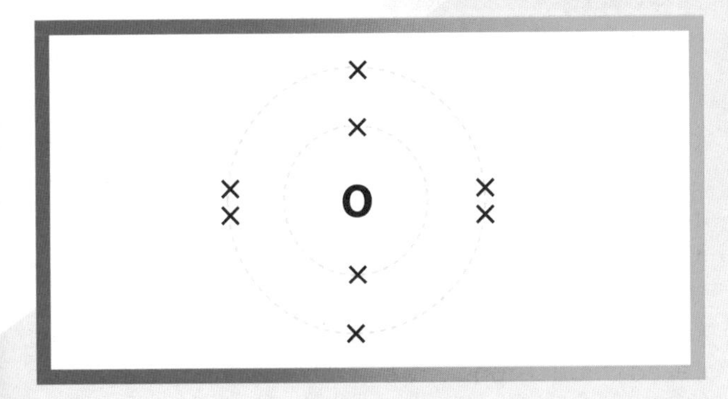

Progress Check

1. Where are electrons found?
2. How many electrons can fit into the first shell?
3. What is the electron configuration of lithium?
4. What group does sodium belong to?
5. What group does carbon belong to?

Exam Question

An atom of magnesium has 12 electrons.

a. What is the electron configuration of magnesium?

b. What group does magnesium belong to?

Ionic Bonding

Ionic compounds are held together by strong forces of attraction between oppositely charged ions, called ionic bonds.

Ions

An **ion** is an atom or group of atoms with a charge.

Ions have a full, outer shell of electrons (like an atom of a noble gas).

Ionic bonding involves the transfer of **electrons** between atoms to form charged ions.

Atoms that gain electrons become negatively charged, while atoms that lose electrons become positively charged.

Ionic **bonding** is the attraction between these oppositely charged ions.

Sodium Chloride

Sodium reacts with chlorine to form sodium chloride. The sodium atom transfers its outer electron to a chlorine atom. Both the sodium atom and the chlorine atom now have a full, outer shell of electrons. The sodium atom which lost an electron now has a 1+ charge and is called a sodium ion, while the chlorine atom has gained an electron and now has a 1– charge and is called a chloride ion.

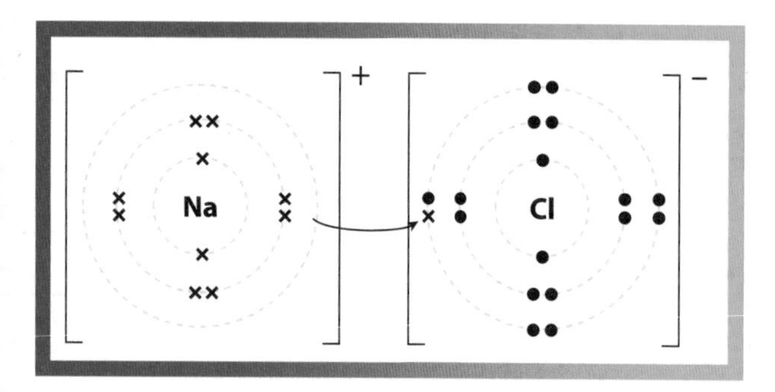

Magnesium Oxide

Magnesium reacts with oxygen to form magnesium oxide. The magnesium atom transfers its two outer electrons to an oxygen atom. The magnesium atom, which lost two electrons, now has a 2+ charge and is called a magnesium ion, while the oxygen atom has gained two electrons and now has a 2– charge and is called an oxide ion.

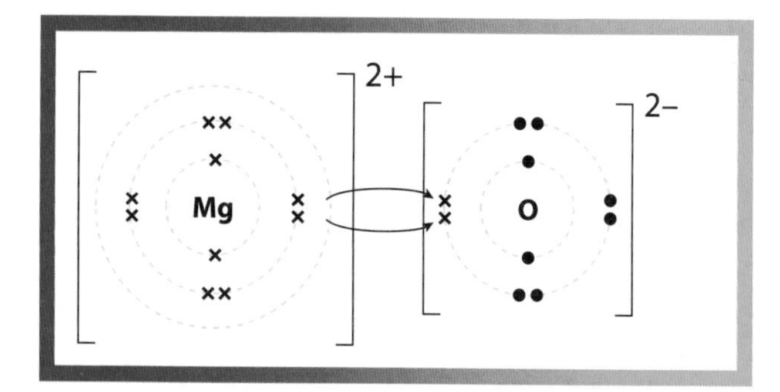

Calcium Chloride

Calcium reacts with chlorine to form calcium chloride. The calcium atom transfers its two outer electrons to two chlorine atoms. The calcium atom, which lost two electrons, now has a 2+ charge and is called a calcium ion, while the chlorine atoms have gained one electron each and now have a 1– charge and are called chloride ions.

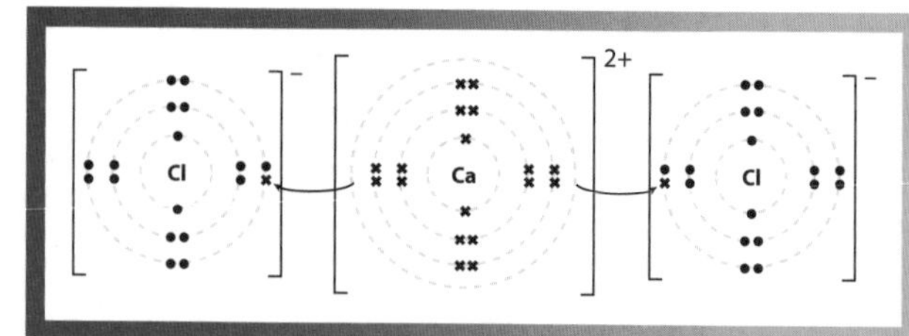

Ionic Structures

Ionic compounds have **giant structures.** They are held together by strong forces of attraction between oppositely charged ions. These forces act in all directions. This means that ionic compounds have very high melting and boiling points because lots of energy must be supplied to break these strong forces of attraction.

Ionic compounds are solid at room temperature. Ionic solids do not conduct electricity because the ions are not able to move, but they do conduct when molten or when they are dissolved in something else.

1. What is an ion?
2. What is the charge on an ion formed when an atom gains one electron?
3. What is the charge on an ion formed when an atom loses one electron?
4. What is the charge on an ion formed when an atom gains two electrons?
5. What is the charge on an ion formed when an atom loses two electrons?

Sodium reacts with chlorine to form an ionic compound.

a. Why does this compound have a high melting point?

b. Write a word equation for the reaction.

c. Sodium atoms lose an electron to form sodium ions. What is the charge on a sodium ion?

Covalent Bonding

In covalent bonding, atoms gain a full outer shell by sharing pairs of electrons.

Covalent Bonds

A covalent bond is a shared pair of **electrons**. Non-metal atoms can gain a fuller, outer shell of electrons by **sharing** pairs of electrons.

Hydrogen, H_2

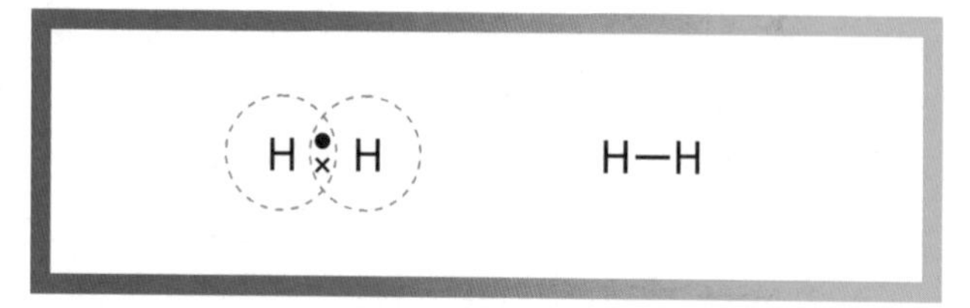

A hydrogen molecule is formed when two hydrogen atoms share a pair of electrons. Both atoms now have a full, outer shell of electrons.

Hydrogen Chloride, HCl

A hydrogen chloride molecule is formed when a hydrogen atom and a chlorine atom share a pair of electrons. Both atoms now have a full, outer shell of electrons.

Methane, CH_4

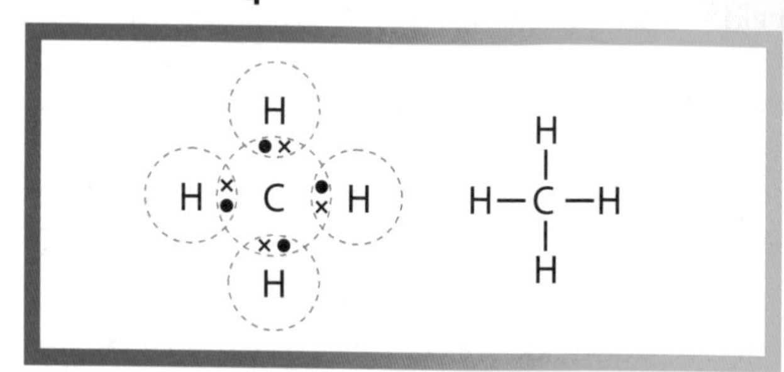

A methane molecule consists of a carbon atom surrounded by four hydrogen atoms. The carbon atom shares a pair of electrons with each of the hydrogen atoms.

Ammonia, NH_3

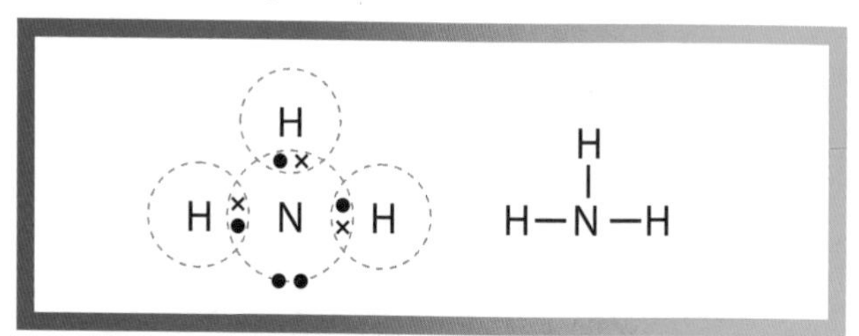

An ammonia molecule consists of a nitrogen atom surrounded by three hydrogen atoms. The nitrogen atom shares a pair of electrons with each of the hydrogen atoms.

Water, H_2O

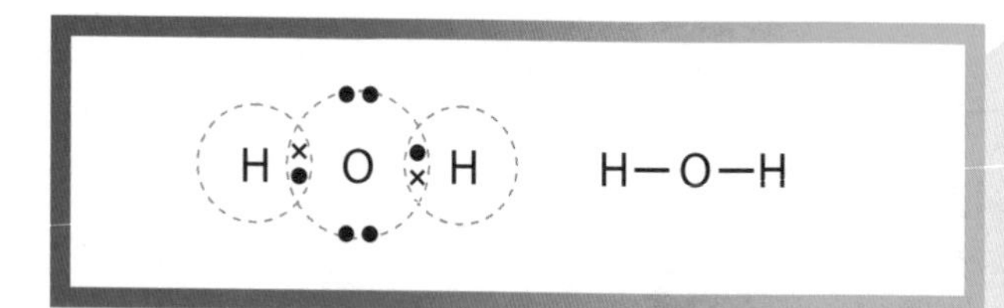

A water molecule consists of an oxygen atom and two hydrogen atoms. The oxygen atom shares a pair of electrons with each of the hydrogen atoms.

Oxygen, O_2

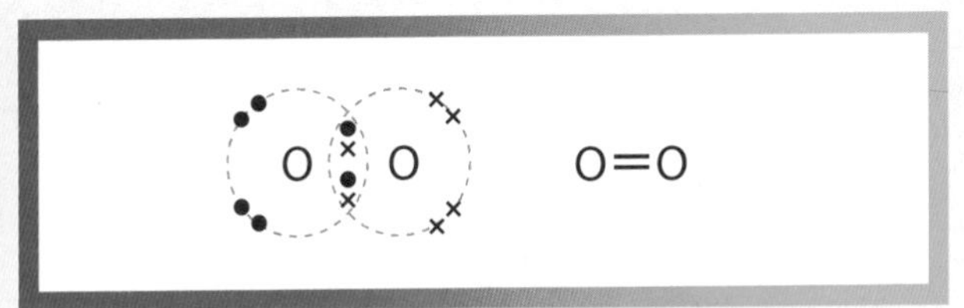

An oxygen molecule consists of two oxygen atoms. The atoms share two pairs of electrons. So they are joined by a double covalent bond.

Types of Structures

There are two types of covalent structure:

- simple covalent structures
- giant covalent structures.

Simple Covalent Structures (Simple Molecules)

Simple covalent structures are formed by small numbers of atoms. There are very strong forces of attraction within these molecules, but much weaker forces of attraction between one molecule and another. This means that simple covalent structures have low melting and boiling points and most are liquids or gases at room temperature. They do not conduct electricity because they do not contain ions or free electrons.

Giant Covalent Structures (Macromolecular)

Giant covalent structures include:

- diamond
- graphite
- silicon dioxide.

They consist of very large numbers of atoms. All the atoms are held together by strong covalent bonds, so they have high melting and boiling points and are solid at room temperature.

They do not conduct electricity (except graphite) because they do not contain ions and they do not have free electrons. They are insoluble in water.

Progress Check

1. Give the formula of a hydrogen molecule.
2. Give the formula of a hydrogen chloride molecule.
3. Give the formula of a methane molecule.
4. Give the formula of an ammonia molecule.
5. Give the formula of an oxygen molecule.

Exam Question

A water molecule consists of an oxygen atom covalently bonded to two hydrogen atoms.

a. What is a covalent bond?

b. What is the formula of water?

c. Why does water have quite a low melting point?

Alkali Metals

The alkali metals belong to **Group 1** of the periodic table. All have similar properties as they have one electron in their outer shell.

The alkali metals include:

- lithium
- sodium
- potassium.

They are found on the far left-hand side of the periodic table.

Group 1 metals have similar properties because they have similar electron structures. Alkali metals react with non-metals to form ionic compounds. The alkali metal atom loses an electron to form an ion with a 1+ charge.

$$Na \rightarrow Na^+ + e^-$$

The alkali metal atom is oxidised.

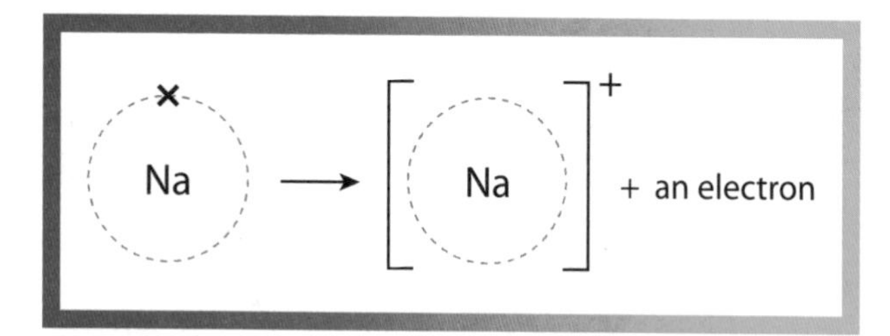

All the alkali metals are very **reactive.** They react vigorously with water to form a metal hydroxide and hydrogen. For this reason they are stored under oil.

Lithium, sodium and potassium are all less dense than water. This means that if a small piece of metal is placed onto a trough of water the metal floats on top of the water as it reacts.

Example

sodium + water → sodium hydroxide + hydrogen

$$2Na + 2H_2O \rightarrow 2NaOH + H_2$$

There is a gradual increase in reactivity down the group. This is because the outer electron is further from the nucleus so it is lost more easily.

Sodium and Chlorine

If sodium is burnt in chlorine gas, the compound sodium chloride is formed.

sodium + chlorine → sodium chloride

The reaction can be summed up as:

$$2Na\,(s) + Cl_2\,(g) \rightarrow 2NaCl\,(s)$$

(s) indicates that the substance is a solid.
(l) indicates that the substance is a liquid.
(g) indicates that the substance is a gas.

Group 1 metals form white compounds that dissolve to form colourless solutions.

Flame Tests

Flame tests can be used to identify the metals in compounds. The colour of the flame indicates the **metal** present:

- lithium – red
- sodium – orange
- potassium – lilac.

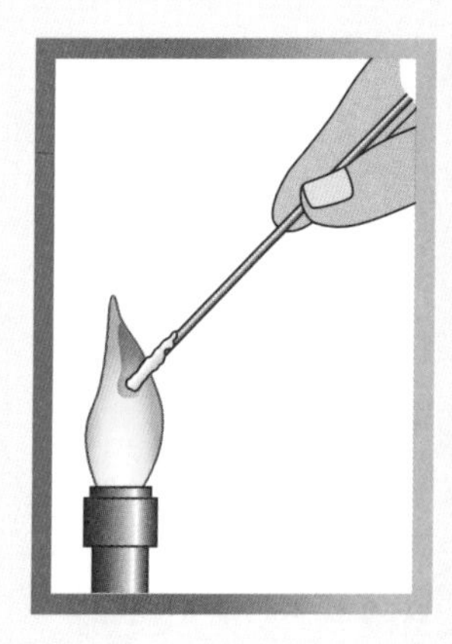

Transition Metals

Transition metals are found in the middle of the periodic table. Compared with Group 1 metals they are hard, strong and much less reactive.

With the exception of mercury, which is a liquid at room temperature, transition metals have higher melting points than Group 1 metals.

While Group 1 metal compounds are white, transition metals form coloured compounds. They are useful catalysts and are also used as pigments and dyes.

1. Where are the alkali metals found in the periodic table?
2. Which of these metals would react most vigorously with water: sodium, lithium or potassium?
3. Name the gas produced when an alkali metal reacts with water.
4. What colour is seen when a flame test is carried out on a sodium compound?
5. What colour is seen when a flame test is carried out on a potassium compound?

EXAM QUESTION

Sodium reacts with water to form sodium hydroxide and hydrogen.

a. Write a word equation for this reaction.

b. Write a balanced symbol equation for this reaction.

c. Why must sodium be stored under oil?

d. When potassium reacts with water the reaction is more vigorous than when sodium reacts with water. Why does potassium react more vigorously?

Noble Gases and Halogens

The noble gases and halogens are found on the right of the periodic table. They have various uses according to their properties.

The horizontal rows in the **periodic table** are called periods. The vertical columns are called groups. Elements in the same group have the same number of electrons in their outer shell, so they have similar chemical properties.

The Noble Gases

The noble gases include:

- helium
- neon
- argon.

They are found on the far right-hand side of the periodic table. All the noble gases are very **unreactive** because they have a full, stable outer shell of electrons. This can make them very useful.

- Helium is **less dense** than air, so if a balloon is filled with helium gas it will float. Helium balloons are popular decorations at parties and special events.
- Argon is used to make filament lamps.
- Neon is widely used in electrical discharge tubes.

The Halogens

The halogens include:

- fluorine
- chlorine
- bromine
- iodine.

They are found next to the noble gases in the periodic table.

Chlorine is a pale green gas, bromine is a brown liquid and iodine is a dark grey solid. This shows us that the **boiling point** of the halogens increases as you go down the group.

Chlorine is used to sterilise water and in the manufacture of pesticides and plastics. Iodine is used to sterilise cuts.

The halogens react vigorously with alkali metals to form metal **halides**. The reaction between sodium and chlorine can be summed up by the equation:

sodium + chlorine → sodium chloride

$2Na + Cl_2 \rightarrow 2NaCl$

The halogen atom, chlorine, gains an electron to form a chloride ion, Cl^-. This is a **reduction** reaction.

Sodium chloride is used in food preparation as a flavouring (common salt) and as a preservative. It is also used in the production of chlorine gas.

There is a gradual decrease in reactivity as you go down the group, so chlorine is more reactive than bromine and iodine. This is because when an atom reacts to form an ion, the electron is placed into a shell increasingly further away from the nucleus.

A more reactive halogen will displace a less reactive halogen from its solution. So chlorine will **displace** bromine from a solution of potassium bromide.

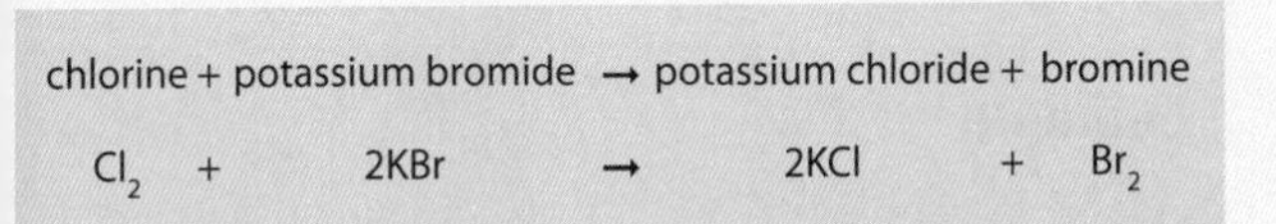

1. Which group does iodine belong to?
2. Which group does neon belong to?
3. In what state is bromine at room temperature?
4. What is helium used for?
5. What is argon used for?

CFCs

Chlorine is used in the manufacture of CFCs. CFCs are chemically inert, have a low boiling point and are insoluble in water. They were widely used as refrigerants and aerosol propellants. If these gases escape into the atmosphere, they can break up the ozone molecules that filter out the Sun's harmful ultraviolet rays.

Consequences of this include:

- a higher risk of sunburn
- skin ages faster
- increased risk of skin cancers and cataracts.

Today CFCs have been replaced by alkanes and HFCs that will not damage the ozone layer.

EXAM QUESTION

Chlorine and bromine both belong to the same group of the periodic table.

a. What group do they belong to?

b. Chlorine is more reactive than bromine. Write a word equation to sum up the reaction between chlorine and potassium bromide.

Calculations 1

Calculations of relative atomic mass and relative formula mass are useful for estimating the amount of product formed in a reaction.

Relative Atomic Mass

Relative atomic mass is used to compare the mass of different atoms. The relative atomic mass of an element is the weighted average mass of the isotopes of the element compared with an atom of carbon-12 which has a mass of 12.

Relative Formula Mass

Relative formula mass is worked out by adding together the relative atomic masses of the atoms in the ratio indicated by the chemical formula.

Example

Water, H_2O

Relative atomic mass of hydrogen = 1

Relative atomic mass of oxygen = 16

The relative formula mass of water = $(1 \times 2) + (16 \times 1) = 18$

Moles

The relative formula mass of a substance in grams is known as one mole of the substance, so one mole of water has a mass of 18 g.

Two moles of water has a mass of 36 g.

Calculating the Mass of a Product

We can work out the theoretical yield of a reaction from a balanced symbol equation.

Example

Hydrogen reacts with oxygen to form water.

$$2H_2 + O_2 \rightarrow 2H_2O$$

If 4 g of hydrogen is burnt, what is the theoretical yield of water produced?

The relative formula mass of $H_2 = 2$

The relative formula mass of $H_2O = 18$

The number of moles in 4 g of $H_2 = \frac{4\,g}{2\,g} = 2$ moles.

From the balanced symbol equation:

2 moles of hydrogen makes 2 moles of water.

The mass of 2 moles of water = $2 \times 18 = 36$ g

So 4 g of hydrogen would form 36 g of water.

Percentage Yield

The amount of product made in a reaction is called the yield. We often find that the actual yield of a reaction is lower than the theoretical yield.

This could be for a number of reasons, including:

- The reaction is reversible and does not go to completion.
- Some of the product was lost, for example during filtering or evaporation.
- There may be side reactions that are producing another product.

The percentage yield of a reaction

$$= \frac{\text{actual amount of product}}{\text{theoretical yield}} \times 100\%$$

Example

The theoretical yield for a reaction is calculated as 1.2 g.

A student carries out the reaction but only produces 1.0 g of product.

The percentage yield $= \frac{1.0}{1.2} \times 100\% = 83.3\%$

Atom Economy

The atom economy of a reaction

$$= \frac{\text{mass of useful product}}{\text{total mass of product}} \times 100\%$$

Scientists try to choose reactions that have a high atom economy.

1. What is relative atomic mass used for?
2. What is the standard against which we measure the relative atomic mass of different atoms?
3. What is one mole equal to?
4. Why might a reversible reaction have a low percentage yield?
5. What is the equation for the percentage yield of a reaction?

EXAM QUESTION

A student calculates that the theoretical yield for her experiment is 2.5 g. She carries out the experiment and only produces 2.2 g of product.

a. Why might her actual yield be less than her theoretical yield?

b. What is the percentage yield of this reaction?

Calculations 2

In chemistry, we use moles to measure the amount of substance.

Moles

The number of moles of a substance

$$= \frac{\text{mass of the substance}}{\text{relative formula mass of the substance.}}$$

This equation can be rearranged to give:

- the relative formula mass of the substance $= \frac{\text{mass of the substance}}{\text{number of moles}}$
- mass of the substance = number of moles × relative formula mass of the substance.

Example

Find the number of moles in 5.85 g of sodium chloride.

The relative formula mass of sodium chloride = 58.5

The number of moles $= \frac{5.85}{58.5} = 0.1$ moles

The number of moles present in solution can also be calculated.

$$\text{Number of moles} = \frac{\text{volume (in cm}^3\text{)}}{1000} \times \text{concentration}$$

This equation can be rearranged to give:

- Volume $= \frac{\text{number of moles} \times 1000}{\text{concentration}}$
- Concentration $= \frac{\text{number of moles} \times 1000}{\text{volume}}$

Volumetric Calculations

We can carry out a **titration** to find the concentration of a solution.

In a titration, a known volume of an alkaline solution is placed into a flask. An **indicator** is added. We can use indicators to find out exactly how much acid is required.

The indicator changes colour when exactly the right amount of acid has been added. The type of indicator used depends on the type of acid and the type of alkali involved.

If we are using a strong acid and a strong alkali then any indicator would work well.

If we are using a strong acid and a weak alkali we should use **methyl orange**.

If we are using a weak acid and a strong alkali we should use **phenolphthalein**.

No indicators work well for a weak acid and a weak alkali.

- Strong acids include hydrochloric acid, nitric acid and sulfuric acid.
- Strong bases include sodium hydroxide and potassium hydroxide.
- Weak acids include carboxylic acids like ethanoic acids.
- Weak bases include ammonium hydroxide or 'ammonia'.

Example

$HCl + NaOH \rightarrow NaCl + H_2O$

20.5 cm^3 of hydrochloric acid, which has a concentration of 0.1 mol dm^{-3}, was required to neutralise 25 cm^3 of sodium hydroxide solution.

What is the concentration of the sodium hydroxide solution?

The number of moles in 20.5 cm^3 of 0.1 mol dm^{-3} hydrochloric acid

$= \frac{20.5}{1000} \times 0.1 = 0.00205$

From the equation 0.00205 moles of hydrochloric acid will react with 0.00205 moles of sodium hydroxide.

The number of moles in 25 cm^3 of alkali
= 0.00205

The concentration of the alkali

$= \frac{\text{number of moles} \times 1000}{\text{volume}}$

$= \frac{0.00205 \times 1000}{25} = 0.082\ \text{mol dm}^{-3}$

The sodium hydroxide solution has a concentration of 0.082 mol dm^{-3}.

1. Carbon dioxide has a formula mass of 44.
   a. How many moles are present in 2.2 g of carbon dioxide?
   b. How many moles are present in 8.8 g of carbon dioxide?
   c. How many moles are present in 11 g of carbon dioxide?
2. A solution of hydrochloric acid has a concentration of 0.1 mol dm^{-3}.
   a. How many moles are present in 25 cm^3 of this solution?
   b. How many moles are present in 100 cm^3 of this solution?

Name the indicator you would use to titrate the weak acid ethanoic acid with the strong alkali sodium hydroxide.

Calculations 3

The empirical formula of a compound is the simplest ratio of atoms in the formula.

Empirical Formula

Ethene has the molecular formula C_2H_4 so it has a empirical formula of CH_2.

We can also use reacting masses to calculate the empirical formula of a compound.

Example

Magnesium reacts with oxygen to form the compound magnesium oxide.

6 g of magnesium reacts with 4 g of oxygen. What is the formula of magnesium oxide?

	Mg	O
Mass	6	4
Find the number of moles (mass/relative atomic mass)…	$\frac{6}{24}$	$\frac{4}{16}$
… this is the ratio in which the atoms combine	0.25	0.25
Divide through by the smallest number…	$\frac{0.25}{0.25}$	$\frac{0.25}{0.25}$
… to get the ratio of atoms in its simplest terms	1	1

The empirical formula of magnesium oxide is MgO.

Bond Enthalpy Calculations

Energy is released when **bonds** are made and taken in when bonds are broken.

Overall, **exothermic** reactions give out energy and **endothermic** reactions take in energy.

Bond energy is the average amount of energy that must be taken in to break one mole of that bond. We can use bond energies to work out whether a reaction is exothermic or endothermic.

Example

Hydrogen reacts with chlorine to form hydrogen chloride. Is this reaction exothermic or endothermic?

Hydrogen + chlorine → hydrogen chloride

$$H_2 + Cl_2 \rightarrow 2HCl$$

	Bond energy (kJ mol^{-1})
H–H	436
Cl–Cl	242
H–Cl	431

The energy taken in to break the old bonds:
1 mole of H–H = 436 kJ
1 mole of Cl–Cl = 242 kJ
Total = 678 kJ

The energy released when the new bonds are made:
2 moles of H–Cl = 2 × 431 = 862 kJ

Overall, more energy is released when the new bonds are formed than is taken in to break the old bonds so the reaction is exothermic.

Avagadro Constant

The Avagadro constant is the number of particles in one mole of a substance. The term 'particles' could refer to atoms, molecules or ions.

Molar Volume

One mole of any gas at a given temperature takes up the same volume.

At a temperature of 25 °C and a pressure of 1 atmosphere one mole of any gas will occupy 24 dm^3.

This means that one mole of nitrogen would occupy a volume of 24 dm^3. Two moles of oxygen would occupy a volume of 48 dm^3. Half a mole of hydrogen would take up a volume of 12 dm^3.

1. What is empirical formula?
2. What is the empirical formula of C_3H_6?
3. What is the empirical formula of N_2H_4?
4. What are the units used to measure bond energy?
5. During a reaction, overall energy is given out. What type of reaction is this?

Methane + oxygen → carbon dioxide and water

$$CH_4 + 2O_2 \rightarrow CO_2 + 2H_2O$$

When one mole of methane is burnt in oxygen, 2644 kJ is required to break the bonds in methane and oxygen.

3338 kJ of energy is released when the bonds in carbon dioxide and water are made.

Is this reaction exothermic or endothermic? Explain your answer.

Haber Process

Ammonia is a very useful raw material in the manufacture of fertilisers and cleaning fluids.

Making Ammonia

The Haber process is used to make ammonia, NH_3. Nitrogen (from the fractional distillation of liquid air) is reacted with hydrogen (from natural gas) in a reversible reaction.

The reaction can be summed up as:

nitrogen + hydrogen ⇌ ammonia

$$N_{2(g)} + 3H_{2(g)} \rightleftharpoons 2NH_{3(g)}$$

The state symbol (g) means gaseous – a gas.

The forward reaction is exothermic.

On cooling, the ammonia liquefies and can be removed. Any unreacted nitrogen and hydrogen are recycled. The ammonia is produced all the time in a continuous process. Ammonia is used in the manufacture of fertilisers and cleaning fluids.

In a closed system, eventually an equilibrium is reached. The relative amount of the substances at equilibrium depends on the conditions.

The conditions chosen in the Haber process are typically:

- an iron catalyst
- a high pressure of around 200 atmospheres
- a moderate temperature of around 450 °C.

An iron catalyst increases the rate of reaction. Catalysts are often used in industry because they allow us to use less energy.

A high pressure is used to increase the yield of ammonia. Increasing the pressure favours the forward reaction, which has fewer gas molecules on the product side.

The forward reaction is exothermic so a high temperature would give a good rate of reaction but a poor yield of ammonia. A low temperature would give a poor rate of reaction but a good yield of ammonia. In practice, a compromise temperature is used that gives a reasonable rate and a reasonable yield.

Fertilisers

Fertilisers help crops to grow bigger and faster. They replace the essential elements including nitrogen, phosphorus and potassium used by plants as they grow. Plants absorb these elements through their roots.

Many fertilisers can be made by a neutralisation reaction between an acid and an alkali.

Ammonia can be oxidised and then reacted with water to form nitric acid. This can be reacted with ammonia to form the popular fertiliser ammonium nitrate.

nitric acid + ammonia → ammonium nitrate

$$HNO_3 + NH_3 \rightarrow NH_4NO_3$$

In a similar way, reacting ammonia with sulfuric acid forms ammonium sulfate, while reacting ammonia with phosphoric acid forms ammonium phosphate.

Percentage by Mass

Plants use nitrogen to produce protein. Being able to calculate the percentage of nitrogen in a fertiliser allows us to choose the right amount of fertiliser to use.

Percentage mass of an element in a compound =

$$\frac{\text{relative atomic mass of the element} \times \text{number of atoms}}{\text{relative formula mass of the compound}} \times 100\%$$

The percentage of nitrogen in ammonium nitrate, NH_4NO_3

$$= \frac{14 \times 2}{80} \times 100\%$$

$= 35\%$

So the percentage of nitrogen in ammonium nitrate is 35%.

Progress Check

1. Where is nitrogen obtained from?
2. Where is hydrogen obtained from?
3. What is the catalyst used in the Haber process?
4. What is the temperature used in the Haber process?
5. What is the pressure used in the Haber process?

Exam Question

Many plant fertilisers contain nitrogen.

a. What do plants use nitrogen for?

b. What is ammonia reacted with to make the fertiliser ammonium nitrate?

c. Calculate the percentage by mass of nitrogen in ammonium nitrate.

The Manufacture of Sulfuric Acid

Some reactions are reversible. If a reversible reaction is carried out in a closed system (where nothing can enter or leave) then eventually a dynamic equilibrium will be reached.

Dynamic Equilibrium

In a **dynamic equilibrium** both the forwards and the reverse reactions are still happening, but they happen at the same rate so that they cancel each other out. Conditions can affect the position of equilibrium. In industrial processes we control the conditions to make the maximum profit.

The Contact Process

The contact process is used to produce sulfuric acid, H_2SO_4.

Stage 1

Sulfur is burnt in oxygen to form sulfur dioxide.

sulfur + oxygen → sulfur dioxide

$$S + O_2 \rightarrow SO_2$$

Stage 2

The sulfur dioxide is reacted with more oxygen over a **vanadium(v) oxide** catalyst at a temperature of around 450 °C and a pressure of 2 atmospheres.

sulfur dioxide + oxygen → sulfur trioxide

$$SO_2 + \frac{1}{2}O_2 \rightleftharpoons SO_3$$

This reaction is reversible and the forward reaction is **exothermic.**

The vanadium(v) oxide catalyst increases the rate of reaction but does not affect the position of equilibrium.

A moderate temperature of around 450 °C is used. A higher temperature would give a better rate of reaction but a poorer yield of sulfur trioxide. A lower temperature would give a better yield of sulfur trioxide but a poorer rate of reaction. A moderate temperature gives a reasonable rate of reaction and a reasonable yield of sulfur trioxide.

A pressure of two atmospheres is used. A higher pressure would give a better yield of sulfur trioxide but the yield is already so high that it is not economically worthwhile to use a very high pressure. A pressure of two atmospheres is enough to keep the gases moving through the system.

Stage 3

The sulfur trioxide cannot be reacted directly with water because the reaction would produce a hazardous fog of sulfuric acid droplets. Instead the sulfur trioxide is reacted with concentrated sulfuric acid, which has a concentration of 98%.

The sulfur trioxide reacts with the water molecules in the acid to form an even more concentrated sulfuric acid. The acid now has a concentration of about 99.5%. This acid is then diluted.

Some acid is retained to produce more sulfuric acid in the future and some is sold to customers.

1. What is a closed system?
2. What is a dynamic equilibrium?
3. What is the catalyst used in the manufacture of sulfuric acid?
4. What is the temperature used in the manufacture of sulfuric acid?
5. What is the pressure used in the manufacture of sulfuric acid?

EXAM QUESTION

In the production of sulfuric acid, sulfur dioxide is reacted with oxygen to form sulfur trioxide.

Sulfur dioxide + oxygen → sulfur trioxide

$$SO_2 + \frac{1}{2}O_2 \rightleftharpoons SO_3$$

a. How is sulfur dioxide produced?

b. What does the symbol $\rightleftharpoons$ mean?

c. A catalyst called vanadium(v) oxide is used in this reaction. How does this catalyst affect the rate of reaction and the position of equilibrium?

Rates of Reaction

The rate of reaction tells us how fast a chemical reaction takes place. A chemical reaction takes place when reacting particles **collide** and have enough **energy** to react – this is called the activation energy.

Increasing the Rate of Reaction

Temperature

When the **temperature** is increased:

- particles move faster
- they collide more often
- when they do collide, they have more energy so more have enough energy to react
- so the rate of reaction increases.

Concentration

When the **concentration** of a solution is increased:

- the particles collide more often
- so the rate of reaction increases.

Concentrations are measured in moles per cubic decimetre, mol dm^{-3}.

Pressure

When the **pressure** of gases is increased:

- the particles collide more often
- so the rate of reaction increases.

Surface Area

When the **surface area** of a solid is increased (small pieces have a large surface area):

- the particles collide more often
- so the rate of reaction increases.

Catalyst

Adding a **catalyst** increases the rate of a chemical reaction. Catalysts are specific to certain reactions. They are not used up during reactions so they can be reused many times. Catalysts are used in industry to reduce production costs. Enzymes are biological catalysts.

Following a Reaction

When magnesium metal is reacted with dilute hydrochloric acid, a salt called magnesium chloride and the gas hydrogen are made.

magnesium + hydrochloric acid → magnesium chloride + hydrogen

We can follow how fast this reaction happens by measuring:

- how quickly the hydrogen gas is made
- how quickly the mass of the reaction flask goes down as the hydrogen gas is made and escapes from the flask.

This graph shows the amount of hydrogen produced in two experiments.

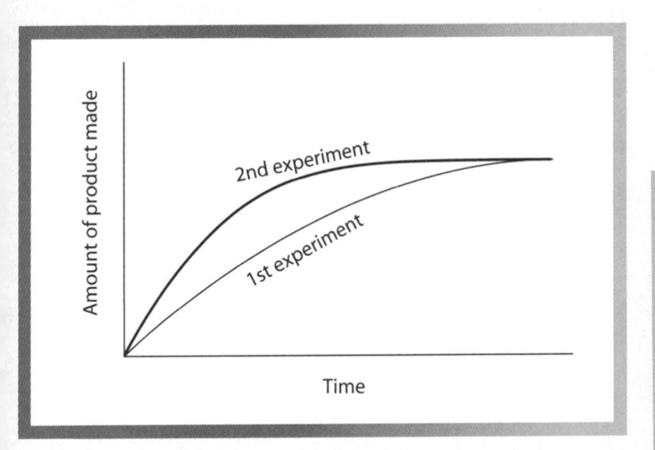

- In both experiments the rate of reaction is fastest at the start of the reaction.
- The reaction is over when the line levels out.
- The second experiment has a faster rate of reaction than the first experiment.

Progress Check

1. What needs to happen for two particles to react?
2. How does increasing temperature affect the rate of a chemical reaction?
3. How does decreasing the concentration of a solution affect the rate of a chemical reaction?
4. How does adding a catalyst affect the rate of reaction?
5. During an experiment, when is the rate of reaction fastest?

Exam Question

Copy and complete the sentences using the words below:

increases **catalysts** **decreases** **reactants**

___a.___ increase the rate of a reaction but they are not used up themselves so they can be used many times. In the reaction between magnesium and hydrochloric acid, increasing the surface area of the magnesium ___b.___ the rate of reaction. Decreasing the concentration of the hydrochloric acid ___c.___ the rate of reaction.

The reaction is over when one of the ___d.___ is used up.

Energy

In all chemical reactions energy is either given out or taken in.

Exothermic Reactions

In **exothermic** reactions, energy (normally in the form of heat) is transferred to the surroundings. This means that if we recorded the temperature change during the reaction, we would see a temperature increase.

The burning of fuels, rusting and neutralisation are all examples of exothermic reactions. Most reactions are exothermic.

The fuel used in Bunsen burners is called methane. When methane is burnt it reacts with oxygen to release heat energy.

Neutralisation reactions are also exothermic. If sodium hydroxide solution is reacted with hydrochloric acid heat energy is released.

Endothermic Reactions

In **endothermic** reactions, energy (normally in the form of heat) is taken in from the surroundings. This means that if we recorded the temperature change during the reaction we would see a temperature decrease.

Thermal decomposition reactions are endothermic.

Bonds

- Energy must be supplied to break **bonds**.
- Energy is released when new bonds are made.
- In exothermic reactions, more energy is released when new bonds are formed than is taken in to break old bonds.
- In endothermic reactions, more energy is taken in to break the old bonds than is released when the new bonds are formed.

The table below shows some average bond energies.

Bond	Bond energy kJ mol^{-1}
C—C	347
O=O	498
C—H	413
O—H	464
C—O	358
H—Cl	432

Reversible Reactions

Some reactions are **reversible**. They can go in a forward or in a reverse direction. If A and B are reactants and C and D are products, then a reversible reaction can be summed up by:

$$A+B \rightleftharpoons C+D$$

If the forward reaction is exothermic, the reverse reaction is endothermic and the same amount of energy is transferred in each case.

When blue, hydrated copper sulfate is heated it decomposes to form white, anhydrous copper sulfate and water. This reaction is endothermic. If water is added to anhydrous copper sulfate it forms hydrated copper sulfate. This reaction is exothermic and can be used to test that a liquid is really water.

The reactions can be summed up by the equation:

hydrated copper sulfate ⇌ anhydrous copper sulfate + water

Ammonia is an important chemical. It is used to make fertilisers. Hydrogen reacts with nitrogen to form ammonia. The reaction is reversible and can be summed up by the reaction:

hydrogen + nitrogen ⇌ ammonia

The forward reaction is exothermic, while the reverse reaction is endothermic.

1. Is the burning of a fuel an exothermic or endothermic reaction?
2. If there is a temperature decrease during a chemical reaction, what type of reaction has taken place?
3. What is released when a new bond is made?
4. What does the symbol ⇌ indicate?
5. What type of reaction is thermal decomposition?

EXAM QUESTION

When hydrated copper sulfate is heated it forms anhydrous copper sulfate and water. The reaction is reversible.

a. Write a word equation for this reaction.

b. What would you **see** when hydrated copper sulfate is heated?

c. How could you prove that a colourless liquid was really water?

Calorimetry

We can measure the energy released when fuels are burned using a technique called calorimetry.

Measuring Energy Changes

In this process:

- We measure the volume of the water in the boiling tube.
- We measure the temperature of the water in the boiling tube.
- We record the mass of the spirit burner and fuel.
- We light the spirit burner and the heat energy released as the fuel burns warms up the water in the boiling tube.
- When 1 g of fuel has been burned we turn out the spirit burner.
- The new temperature of the water is measured.

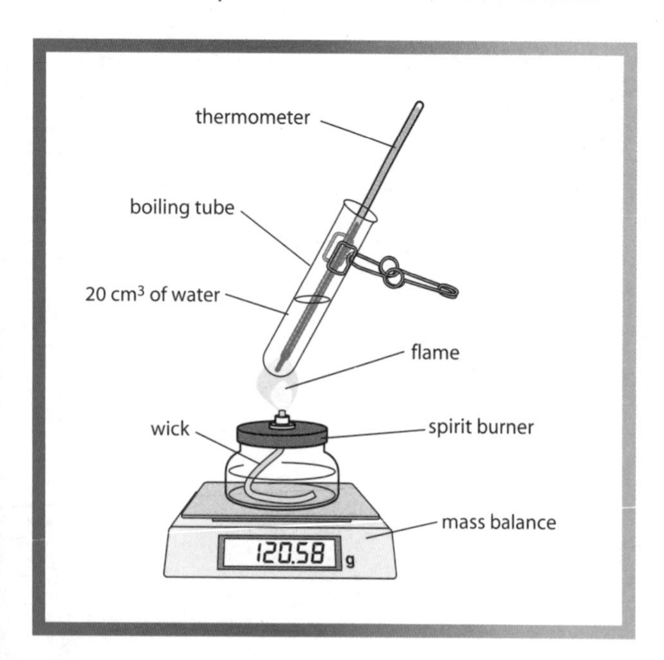

Calculating the Energy Transferred

We can calculate the heat energy transferred using the equation:

Heat energy transferred (J)
= mass of water (g)
× specific heat capacity of water ($J\ g^{-1}\ °C^{-1}$)
× change in temperature (°C)

The specific heat capacity of water is $4.2\ J\ g^{-1}\ °C^{-1}$

This means that it takes 4.2 J of energy to raise the temperature of 1g of water by 1 °C.

In a calorimetry experiment 1g of fuel raised the temperature of 10g of water by 5°C.

The heat energy transferred is:

Heat energy transferred (J) = $10\ g \times 4.2 J\ g^{-1}\ °C^{-1} \times 5°C$
= 210 J

To compare the amount of energy released by different fuels we can divide the heat energy transferred by the mass of fuel burned.

Energy in Fuels

The table below shows the amount of energy released when 1 g of each fuel was burned.

Fuel	Energy released (kJ)
hydrogen	143
petrol	48
ethanol	30

Hydrogen is a gaseous fuel. It releases most energy per gram, but because it is a gas large volumes of hydrogen are required.

Petrol is a liquid but is a non-renewable fuel.

Ethanol is also a liquid but releases less energy per gram than petrol. It is a renewable fuel.

Energy in Foods

The labels on food display nutritional information. They tell us the amount of energy that the food contains. This is important because it helps us to make sure we get the right amount of energy from the food we eat. If we consumed either too much energy or not enough energy we could become ill.

In science we normally measure the amount of energy in joules. For historical reasons we normally measure the energy in foods in calories. One calorie is equal to 4.2 joules.

Progress Check

1. What is the name of the technique used to measure the amount of energy released when fuels are burnt.
2. What is the unit used to measure the amount of heat energy transferred?
3. What is the unit used to measure the change in temperature?
4. What is the unit used to measure the mass of water?
5. What is one calorie equal to?

Exam Question

In a calorimetry experiment, 1g of fuel raised the temperature of 5g of water by 4 °C.

a. Is the burning of this fuel an exothermic or an endothermic reaction?

b. Calculate the heat energy transferred during this reaction.

Energy Profile Diagrams

We use energy profile diagrams to show the energy changes that happen in exothermic and endothermic reactions.

Exothermic Reactions

In exothermic reactions, the products contain less energy than the reactants, so energy is released by the reaction.

The activation energy is the amount of energy required to break the bonds in the reactants and get the reaction started.

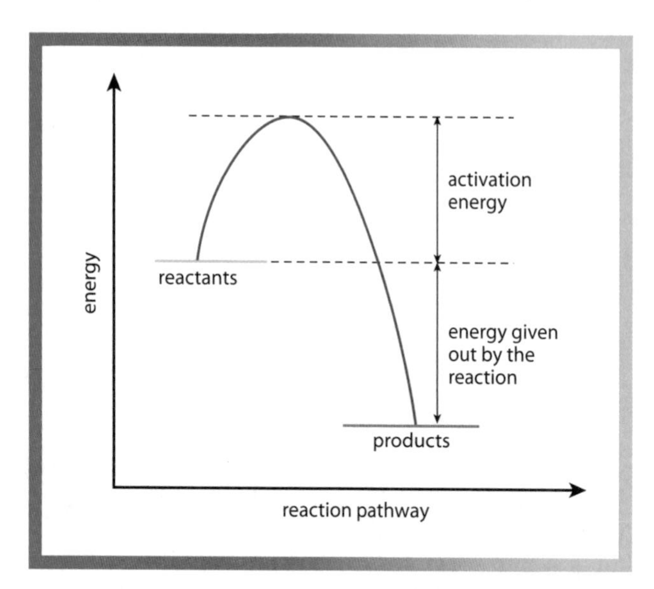

The burning of fuels, neutralisation and rusting are all exothermic reactions.

Endothermic Reactions

In endothermic reactions, the products contain more energy than the reactants so energy is taken in from the surroundings.

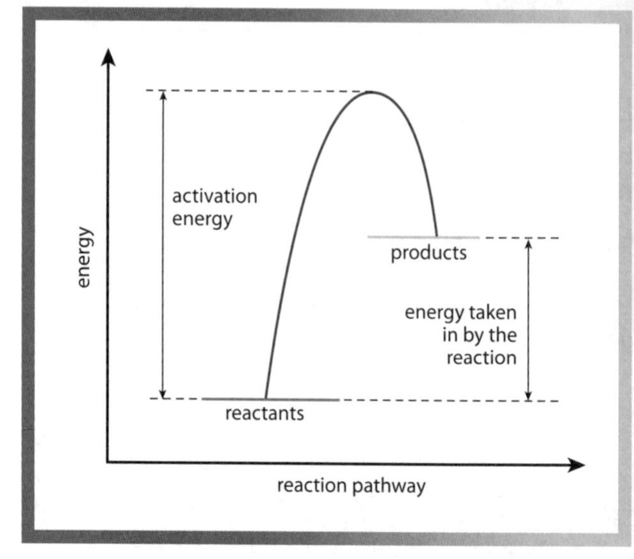

The thermal decomposition of calcium carbonate is an exothermic reaction.

Catalysts

Catalysts are substances that speed up the rate of a chemical reaction but are not used up themselves. Catalysts work by offering an alternative reaction pathway with lower activation energy.

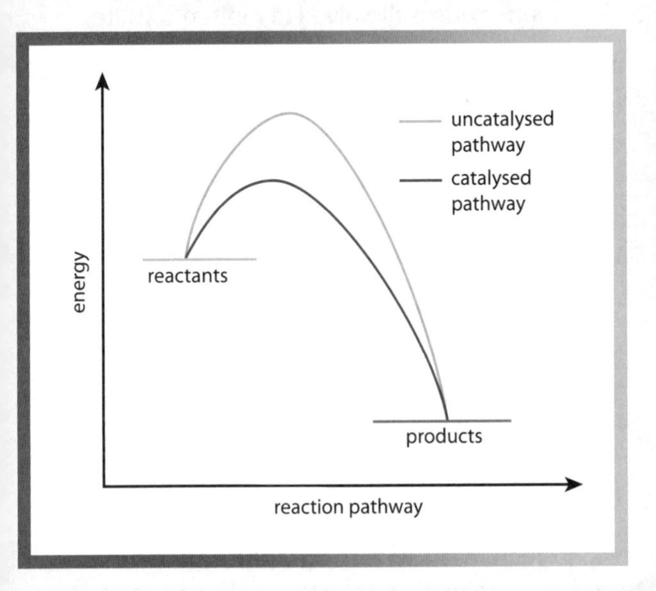

PROGRESS CHECK

1. What is taken in when bonds are broken?
2. What is the activation energy?
3. What is given out during an exothermic reaction?
4. What do catalysts do?
5. Why can catalysts be reused?

EXAM QUESTION

Iron is used as a catalyst in the Haber process, which is used to make ammonia.

$N_2 + 3H_2 \rightleftharpoons 2NH_3$

The forward reaction is exothermic.

a. Draw an energy profile diagram for the forward reaction.

b. Describe, in terms of activation energy, how iron increases the rate of the reaction.

Aluminium

Aluminium is extracted from its ore by electrolysis and has many uses.

Pure aluminium has a low **density** but is too soft for many uses. Aluminium can be mixed with other metals to form **alloys**. These alloys combine low density with high strength.

Extraction

Aluminium alloys combine high strength with low density

Aluminium is quite a reactive metal. It is extracted from its ore, **bauxite**, by **electrolysis**. This is an expensive process because it involves lots of steps and requires lots of energy.

Bauxite contains aluminium oxide. For electrolysis to occur, the aluminium ions and the oxide ions must be able to move. So solid aluminium oxide must either be heated until it melts or dissolved in something else.

Aluminium oxide has a very high melting point, so heating aluminium oxide until it melts would be very expensive. Fortunately, another aluminium ore called **cryolite** has a much lower melting point. In practice, aluminium oxide is dissolved in molten cryolite.

During electrolysis the Al^{3+} ions move to the negative electrode (cathode) where aluminium forms.

$$Al^{3+} + 3e^- \rightarrow Al$$

The aluminium ions are **reduced**.

The oxide ions, O^{2-}, move to the positive electrode (anode) where they react to form oxygen molecules.

$$2O^{2-} \rightarrow O_2 + 4e^-$$

The oxide ions are **oxidised**.

The oxygen reacts with the **graphite** anode to produce carbon dioxide so these electrodes must be regularly replaced.

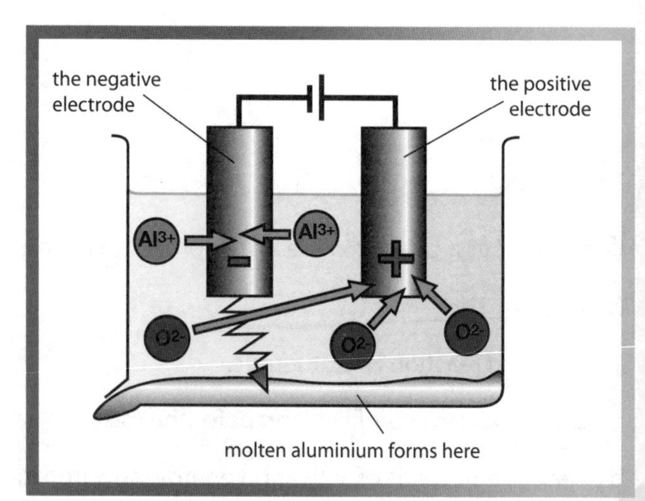

The Electrolysis of Dilute Sulfuric Acid

The electrolysis of dilute sulfuric acid produces **oxygen** and **hydrogen.**

Hydrogen is formed at the cathode.

$$2H^+ + 2e^- \rightarrow H_2$$

Oxygen is formed at the anode.

$$4OH^- \rightarrow 2H_2O + O_2 + 4e^-$$

Gas Tests

Hydrogen	Oxygen
Burns with a 'squeaky' pop	Relights a glowing splint

1. What is the chemical compound found in bauxite?
2. Bauxite is one ore of aluminium. Name another ore of aluminium.
3. Does solid aluminium oxide conduct electricity?
4. Does molten aluminium oxide conduct electricity?
5. In the electrolysis of aluminium oxide, what is formed at the negative electrode?

EXAM QUESTION

Aluminium is extracted from aluminium oxide by electrolysis.

a. During electrolysis, aluminium is deposited at the negative electrode. Give the symbol equation for this reaction.

b. Oxygen is produced at the positive electrode. Give the symbol equation for this reaction.

c. Why must the graphite electrodes be regularly replaced?

Sodium Chloride

Sodium chloride is a salt. Salts are very important compounds with many uses.

Uses of salts include:

- the production of fertilisers
- as colouring agents
- in fireworks.

Sodium Chloride Solution

Sodium chloride (common salt) is a very important resource. It is found in large quantities dissolved in seawater and in underground deposits formed when ancient seas evaporated.

Rock salt (unpurified salt) is used during winter to grit roads to stop them from becoming icy and dangerous.

The **electrolysis** of concentrated sodium chloride solution produces:

- chlorine
- hydrogen
- sodium hydroxide.

During electrolysis, positive hydrogen H^+ ions move to the negative electrode (cathode) where hydrogen is produced.

$$2H^+ + 2e^- \rightarrow H_2$$

Chloride ions, Cl^-, move to the positive electrode (anode) where chlorine is produced.

$$2Cl^- \rightarrow Cl_2 + 2e^-$$

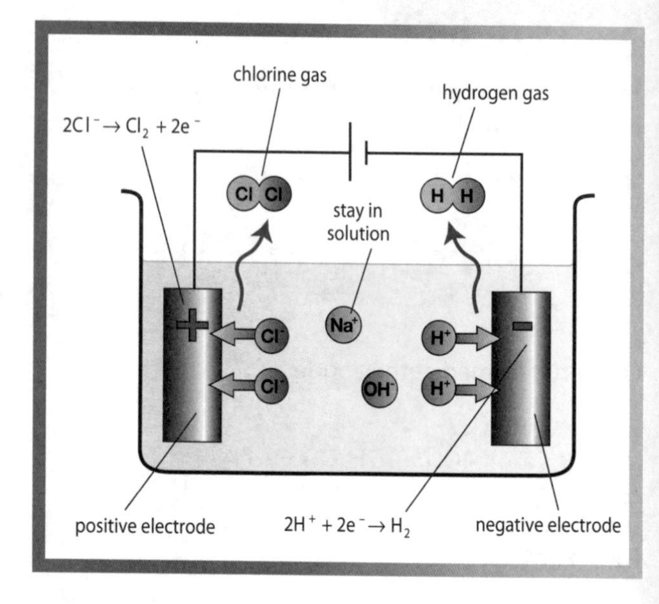

Gas tests:

- Hydrogen burns with a 'squeaky' pop.
- Chlorine bleaches damp litmus paper.

Uses

The products of the electrolysis of concentrated sodium chloride are very important.

- Chorine is used to sterilise water.
- Hydrogen is used in the manufacture of margarine.
- Sodium hydroxide is also produced and is used in the production of soaps.

Solid Sodium Chloride

Solid sodium chloride does not conduct electricity because the ions cannot move. If sodium chloride is heated until it melts, then electrolysis can occur.

The electrolysis of molten sodium chloride produces sodium and chlorine.

Chloride ions, Cl^- move to the positive electrode (anode) where chlorine is produced.

$$2Cl^- \rightarrow Cl_2 + 2e^-$$

Positive sodium Na^+ ions move to the negative electrode (cathode) where sodium is produced.

$$Na^+ + e^- \rightarrow Na$$

Insoluble Salts

Some salts are **insoluble.** They can be made by reacting solutions of soluble salts.

Here, insoluble barium sulfate is made by reacting a solution of barium chloride with a solution of sodium sulfate. The reaction also produces sodium chloride.

The barium sulfate is a **precipitate.**

barium chloride + sodium sulfate → barium sulfate + sodium chloride

$$BaCl_{2\,(aq)} + Na_2SO_{4\,(aq)} \rightarrow BaSO_{4\,(s)} + 2NaCl_{(aq)}$$

The state symbol (aq) means aqueous (it is dissolved in water). The state symbol (s) means it is solid.

Progress Check

1. Give **one** use of salts.
2. Name an insoluble salt
3. Give the word equation for the reaction between barium chloride and sodium sulfate.
4. What does the state symbol (s) indicate?
5. What does the state symbol (aq) indicate?

Exam Question

Sodium chloride is an important resource.

a. The electrolysis of concentrated sodium chloride solution produces two gases. Chlorine is produced at the positive electrode. Which gas is produced at the negative electrode?

b. Give the equation for the reaction at the positive electrode that produces chlorine.

c. Give **one** use of chlorine.

Copper

Copper is widely used for electrical wiring and plumbing.

Uses

Copper is a very useful metal, it is:

- a good electrical conductor
- a good thermal conductor
- very unreactive.

Extraction

Copper is a fairly unreactive metal. It can be extracted from its ore simply by heating. Copper produced in this way may contain some impurities.

Copper can be purified by electrolysis.

Electrolysis

In electrolysis the impure copper is used as the positive electrode (anode).

Pure copper is used as the negative electrode (cathode).

A solution that contains Cu^{2+} ions is also used.

At the anode, copper atoms give up electrons to form copper ions. These ions dissolve into solution, so the anode loses mass.

$$Cu\ (s) \rightarrow Cu^{2+}(aq) + 2e^-$$

At the cathode, copper ions gain electrons to form copper atoms. Over time the cathode gets coated in very pure copper, so the cathode gains mass.

$$Cu^{2+}(aq) + 2e^- \rightarrow Cu\ (s)$$

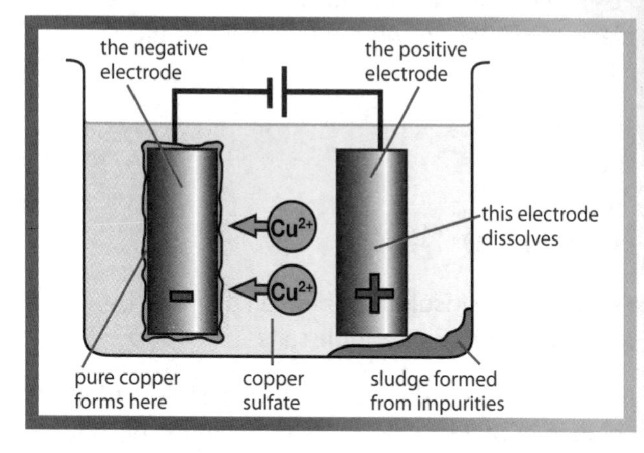

- The loss in mass of the anode is equal to the gain in mass of the cathode.
- The higher the electrical current the more copper is produced at the cathode.
- The longer the time the more copper is produced at the cathode.
- The relationship between the charge transfer, current and charge is given by:

$$\text{charge} = \text{current} \times \text{time}$$

Electrolysis of Ionic Compounds

Ionic substances contain atoms or groups of atoms that have a charge. The ions in solid ionic substances have fixed positions but if the substances are dissolved or heated until they melt, the ions can move.

The electrolysis of molten aluminium oxide, Al_2O_3 produces aluminium and oxygen.

At the anode:

$$2O^{2-} \rightarrow O_2 + 4e^-$$

At the cathode:

$$Al^{3+} + 3e^- \rightarrow Al$$

The electrolysis of molten lead bromide, $PbBr_2$ produces lead and bromine.

At the anode:

$$2Br^- \rightarrow Br_2 + 2e^-$$

At the cathode:

$$Pb^{2+} + 2e^- \rightarrow Pb$$

The electrolysis of molten lead iodide, PbI_2 produces lead and iodine.

At the anode:

$$2I^- \rightarrow I_2 + 2e^-$$

At the cathode:

$$Pb^{2+} + 2e^- \rightarrow Pb$$

The electrolysis of molten potassium chloride, KCl, produces potassium and chlorine.

At the anode:

$$2Cl^- \rightarrow Cl_2 + 2e^-$$

At the cathode:

$$K^+ + e^- \rightarrow K$$

Progress Check

1. How can copper be extracted from its ore?
2. How is copper purified?
3. What is the positive electrode called?
4. What is the negative electrode called?
5. During the purification of copper by electrolysis, what type of ion does the solution contain?

Exam Question

Copper can be purified by electrolysis. Impure copper is placed at the positive electrode (anode).

a. Give the symbol equation for the reaction that takes place at the positive electrode.

b. Copper atoms are deposited at the negative electrode. Give the symbol equation for the reaction that takes place at the negative electrode.

Acids and Bases

In 1883, the Swedish scientist Svante Arrhenius suggested that acids produce **H^+ ions** and bases produce **OH^- ions**.

History

Arrhenius' explanation works well in aqueous solutions but because it does not work all the time, it was not generally accepted at first.

It does not explain why some compounds that do contain hydrogen do not form acidic solutions or why some compounds, like sodium carbonate, that do not contain hydroxide ions are still bases.

These problems were addressed in 1923 by Johannes Bronsted and Thomas Lowry. These two scientists worked separately but both proposed that acids are proton, H^+ donors and bases are proton, H^+ acceptors.

This new definition extends Arrhenius' earlier model allowing it to be used in many more situations.

It also explains the importance of water. Water, H_2O, accepts H^+ ions to form H_3O^+ ions.

It also works in situations where solvents other than water are used.

Many foods contain weak acids

Strength of Acids and Bases

We describe the strength of an acid or a base by the extent to which it is ionised in water.

Strong acids like hydrochloric acid, sulfuric acid and nitric acid are completely ionised in water. Strong alkalis like sodium hydroxide and potassium hydroxide are also completely ionised in water.

Weak acids like ethanoic acid, citric acid and carbonic acid are only partially ionised in water.

Weak alkalis like ammonia are also only partially ionised in water.

Many cleaning materials contain bases

The amount of ionisation affects conductivity.

- Hydrochloric acid is a strong acid, it completely ionises in water so it has a high electrical conductivity.

$$HCl \rightarrow H^+ + Cl^-$$

- Ethanoic acid is a weak acid, it only partially ionises in water so it has a much lower electrical conductivity.

$$CH_3COOH \rightleftharpoons CH_3COO^- + H^+$$

Electrolysis of both of these acids produces hydrogen gas at the negative electrode.

Concentration

The concentration of an acid or an alkali is the number of moles in 1dm^3.

If the same concentration of hydrochloric acid and ethanoic acid were reacted with magnesium, the ethanoic acid would react more slowly than the hydrochloric acid.

This is because the hydrochloric acid would have a higher concentration of hydrogen ions, so there would be a greater collision frequency between magnesium and hydrogen ions and the reaction would happen faster.

PROGRESS CHECK

1. What does a proton consist of?
2. Write the formula of a hydroxide ion.
3. What is formed when a water molecule joins with a proton?
4. Name **three** strong acids.
5. Name a weak alkali.

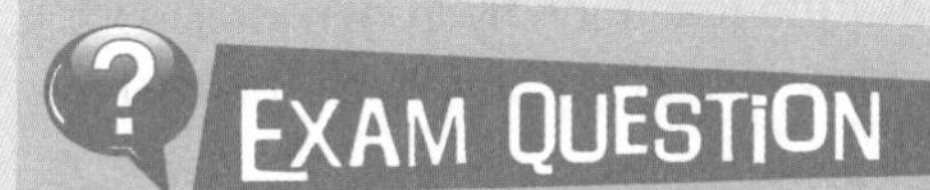

Acids can be described as strong or weak.

a. Name a weak acid.

b. What is the difference between a strong acid and a weak acid?

Making Salts

Salts are very important compounds. They can be made by reacting acids with metals, metal carbonates, metal oxides or metal hydroxides.

- Hydrochloric acid forms **chloride** salts.
- Nitric acid forms **nitrate** salts.
- Sulfuric acid forms **sulfate** salts.

Sulfuric acid is used in the manufacture of fertilisers and in car batteries.

Solutions

If a solution has a pH of 7 it is **neutral**. If the pH is less than 7 it is an **acid** and if the pH is more than 7 it is an **alkali**.

Acids react with bases to form a salt and water.

The reaction between an acid and a base is called **neutralisation**.

An alkali is a soluble base.

Acidic solutions contain **hydrogen**, H^+ ions. Alkaline solutions contain **hydroxide**, OH^- ions.

During **neutralisation** reactions, hydrogen ions react with hydroxide ions to form water, which is neutral.

$$H^+_{(aq)} + OH^-_{(aq)} \rightarrow H_2O_{(l)}$$

The state symbol (l) means it is a liquid.
The state symbol (aq) means that it is aqueous or dissolved in water.

Metals

Fairly reactive metals react with acids to form a salt and hydrogen.

magnesium + hydrochloric acid → magnesium chloride + hydrogen

$$Mg + 2HCl \rightarrow MgCl_2 + H_2$$

Metal Carbonates

Metal carbonates react with acids to form a salt, water and carbon dioxide.

calcium carbonate + hydrochloric acid → calcium chloride + water + carbon dioxide

$$CaCO_3 + 2HCl \rightarrow CaCl_2 + H_2O + CO_2$$

The metal carbonate is added to the acid until the reaction stops (there is no more fizzing). The excess metal carbonate is then filtered off leaving a salt solution. If the salt solution is warmed carefully, the water evaporates leaving crystals of the salt.

Metal Oxides

Metal oxides react with acids to form a salt and water.

copper oxide + nitric acid → copper nitrate + water

$$CuO + 2HNO_3 \rightarrow Cu(NO_3)_2 + H_2O$$

Metal Hydroxides

Metal hydroxides also react with acids to form a salt and water.

potassium hydroxide + nitric acid → potassium nitrate + water

$KOH + HNO_3 \rightarrow KNO_3 + H_2O$

An indicator can be used to show when the acid and alkali have completely reacted.

Ammonia

Ammonia can be reacted with acids to form ammonium salts.

ammonia + hydrochloric acid → ammonium chloride

$NH_3 + HCl \rightarrow NH_4Cl$

1. If a solution has a pH of 7, what type of solution is it?
2. If a solution has a pH of 10, what type of solution is it?
3. If a solution has a pH of 2, what type of solution is it?
4. Name the salt made when magnesium reacts with hydrochloric acid.
5. Name the salt made when calcium carbonate reacts with hydrochloric acid.

Exam Question

Egg shells contain calcium carbonate. If hydrochloric acid is placed on an egg shell, the acid reacts with the calcium carbonate.

a. Write a word equation to sum up this reaction.

b. Suggest a value for the pH of hydrochloric acid.

Metals

Metals all have similar properties because of their atomic structure. Metals have many uses.

Metallic Bonding

The atoms in a pure metal have a regular arrangement. This means that the atoms can slide over each other quite easily so metals can be bent into shape, or drawn into wires.

Metals can be mixed to form alloys. In alloys, the atoms have different sizes so it is harder for the layers of atoms to pass over each other. This makes alloys harder than pure metals.

In metals, the **electrons** in the outermost shell of an atom are **delocalised** and free to move throughout the whole structure. This means that metals consist of positive metal ions and negative delocalised electrons.

Metallic bonding is the attraction between these positive metal ions and the negative delocalised electrons. This is an **electrostatic** attraction.

Metals are good thermal and electrical conductors because the delocalised electrons are free to move throughout the structure.

Metals have high melting points because of the strong forces of attraction between the metal ions and the delocalised electrons.

Metal Carbonates

When metal **carbonates** are heated, they decompose to form a metal oxide and carbon dioxide. This is known as a thermal decomposition reaction. If carbon dioxide is bubbled through limewater it turns the limewater cloudy.

copper carbonate	→	copper oxide	+	carbon dioxide
$CuCO_3$	→	CuO	+	CO_2
iron carbonate	→	iron oxide	+	carbon dioxide
$FeCO_3$	→	FeO	+	CO_2
manganese carbonate	→	manganese oxide	+	carbon dioxide
$MnCO_3$	→	MnO	+	CO_2
zinc carbonate	→	zinc oxide	+	carbon dioxide
$ZnCO_3$	→	ZnO	+	CO_2

We can tell that a chemical reaction has taken place when there is a change of colour, which shows us that a new substance has been formed.

Hydroxide Tests

We can identify the metals present in metal salt solutions by adding sodium **hydroxide solution**.

If the metal ion forms a **precipitate**, we can use the colour of the precipitate to identify the metal present:

- copper(II) – pale blue precipitate
- iron(II) – green precipitate
- iron(III) – brown precipitate.

Transition Metals

Transition metals are found in the middle of the periodic table.

Many transition metals and their compounds are useful catalysts.

Iron is used in the manufacture of ammonia and nickel is used in the manufacture of margarine.

Most iron is made into the alloy **steel**, which is used to make cars and bridges because it is strong and cheap.

Copper is used to make **brass**, which is used in electrical wiring because it is a good electrical conductor.

Superconductors

Some metals behave as **superconductors** at very low temperatures.

Superconductors are special because when they conduct electricity, they have very little or no resistance. Use of superconductors will be limited until we can develop types that will work at room temperature.

Progress Check

1. What is metallic bonding?
2. What are mixtures of metals called?
3. What is the catalyst used in the manufacture of ammonia?
4. What is the catalyst used in the manufacture of margarine?
5. What is the test for carbon dioxide?

Transition metals are strong and hard.

a. Name an alloy of iron used to make cars.

b. Name an alloy of copper used in electrical wiring.

Useful Metals

Pure aluminium has a low **density** but is too soft for many uses.

Aluminium

Aluminium can be mixed with other metals to form **alloys**. These alloys combine low density with high strength.

Other common alloys are:

- amalgram (mainly mercury)
- **brass** (copper and zinc)
- solder (lead and tin)
- bronze (copper and tin).

Aluminium is a reactive metal. It is extracted from its ore, bauxite, by **electrolysis**. This is an expensive process because it involves many steps and requires lots of energy.

Aluminium is actually much more reactive than it appears. This is because aluminium objects quickly react with oxygen to form a layer of aluminium oxide, which prevents any further reaction from occurring.

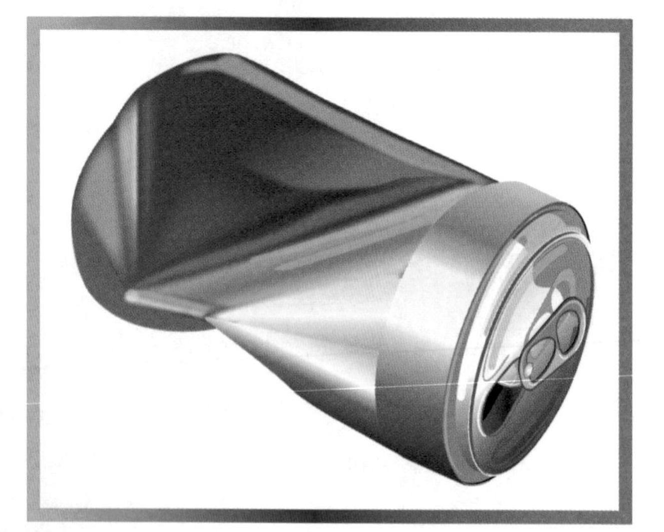

Aluminium can be used to make many things, including:

- drinks cans
- bicycles
- aeroplanes.

Aluminium Cars

Car bodies are normally made from steel, but they can also be made from aluminium.

The advantages of using aluminium are as follows:

- The car body will be lighter so the car will have a better fuel economy.
- The car body will corrode less so it may last for longer.

One of the disadvantages of using aluminium is that an aluminium car body will be more expensive to produce.

Protection of the Environment

The advantages of **recycling** old aluminium objects are as follows:

- Landfill sites are not filled up as quickly.
- Recycling old aluminium objects uses much less energy than extracting aluminium from its ore.
- Recycling old aluminium objects means that less aluminium ore needs to be extracted, which protects the environment.

Titanium

Titanium is a very useful metal:

- It has a low density.
- It is strong.
- It is very resistant to corrosion.
- It has a very high melting point.

Titanium is a reactive metal, so extracting it from its ore, rutile, is a difficult process.

Copper

Copper is a very useful metal, which is:

- a good electrical conductor
- a good thermal conductor
- very resistant to corrosion
- very unreactive.

Copper is widely used for electrical wiring and plumbing. Traditionally, we extract copper from its ores and then purify it using electrolysis. Today we have to extract copper from ores that actually contain very little copper. This means that very large quantities of rock must be quarried which cause environmental problems. Scientists are developing ways to extract copper from low grade ores to try to minimise these environmental problems.

1. Why is pure aluminium not widely used?
2. What is a mixture of metals called?
3. What is brass made from?
4. Give one use of aluminium.
5. What is the ore of titanium called?

EXAM QUESTION

Aluminium can be used to make car bodies.

a. Why does aluminium appear to be less reactive than it really is?

b. Give one advantage of using aluminium rather than steel to make car bodies.

c. Give one disadvantage of using aluminium rather than steel to make car bodies.

Iron and Steel

Gold is a very unreactive metal. It is found in nature uncombined. Iron is a fairly reactive metal. It is extracted from its ore, which contains iron(III) oxide, in the **blast furnace**.

Extraction of Iron

An **ore** is a rock which contains metals in such high concentrations that it is economically worthwhile to extract the metal from the rock.

Iron is less reactive than carbon. To extract iron from iron (III) oxide, the oxygen must be removed. This reaction is called reduction. Iron produced by the blast furnace contains quite large amounts of carbon. If it is allowed to cool down and solidify, **cast iron** is produced. Cast iron contains around 96% pure iron.

Cast iron is:

- very hard
- strong
- resistant to corrosion
- very brittle.

Wrought Iron

Wrought iron is produced by removing the carbon from cast iron. Because it is very pure the iron, atoms form a very regular arrangement. This means that the layers of iron atoms can slip over each other very easily, which makes wrought iron very easy to shape. Unfortunately, it can be too soft for many uses.

Steel

Most iron is made into **steel**. First the carbon impurities are removed, then carefully controlled amounts of carbon and metals like chromium are added.

We can produce steels that have quite different properties.

- Low carbon steels are soft and easy to shape.
- High carbon steels are hard but brittle.
- **Stainless steel** is very resistant to corrosion.

Steel is harder than wrought iron because it contains atoms of iron, carbon and other metals that are different sizes. This means the atoms cannot form a regular arrangement. This makes it very difficult for the layers to slip over each other, which makes steel hard.

Steels are also less likely to corrode than pure iron.

Rusting

The corrosion of iron is called rusting. During rusting, iron reacts with water and oxygen to form hydrated iron(III) oxide.

This can be summarised by the word equation:

iron + water + oxygen → hydrated iron(III) oxide

Salt increases the rate of rusting, so cars at the seaside may rust faster than cars away from the coast.

1. Name the iron compound found in iron ore.
2. Name a metal that is found uncombined in nature.
3. Where is iron extracted?
4. Iron is made by removing oxygen from iron oxide. What is this type of reaction called?
5. Roughly, what is the percentage of iron in cast iron?

EXAM QUESTION

Use the words below to complete the table.

cast iron   wrought iron   low carbon steel   iron(III) oxide

Name	Description
a.	The name of very pure iron
b.	An alloy containing iron and carbon, which is easy to shape
c.	A material produced when iron from the blast furnace is allowed to cool down and solidify
d.	The compound found in iron ore

Water

Water is a very important resource, which needs to be conserved.

Water is used as:

- a coolant
- a raw material to make new chemicals from
- a solvent.

Water Pollution

Water can be polluted by:

- nitrates from fertiliser run off
- lead compounds from lead pipes
- pesticides sprayed near to water resources.

Nitrate Problems

Fertilisers are used to help plants to grow better, but they can enter waterways and cause problems.

The nitrates in fertilisers can cause algae to grow at a fast rate. Eventually the algae die and bacteria begin to decompose them. The bacteria use up the oxygen in the water. Fish and other aquatic life cannot get enough oxygen and may die. This is called eutrophication.

In the UK, water is taken from areas that are well away from any sources of pollution, including:

- lakes
- rivers
- aquifers (rocks that contain water)
- reservoirs.

There is only a limited supply of fresh water, so it is important that we conserve water when we can.

Water Purification

In developing nations clean, safe water is extremely important.

In the water purification process:

- any large pieces of debris such as sticks are removed
- the water is filtered to remove suspended particles such as clay
- the water is passed through filter beds of charcoal and sand to remove unpleasant smells and make the water taste better
- finally, the water is chlorinated to reduce the number of micro-organisms to acceptable levels.

Water is only taken from safe sources

Water can be obtained from seawater by distillation but this takes a lot of energy and is only used where fuel is very cheaply available.

Water Filters

The taste and quality of tap water can be improved by filtering the water. These filters contain carbon, silver and ion exchange resins.

Insoluble Salts

Some salts are **insoluble**. They can be made by reacting solutions of soluble salts.

We can use a solution of silver nitrate to test for chloride, bromide or iodide ions.

- If silver nitrate is added to a solution that contains chloride ions, we see a white precipitate of silver chloride.
- If it is added to a solution that contains bromide ions, we see a cream precipitate of silver bromide.
- If it is added to a solution that contains iodide ions, we see a pale yellow precipitate of silver iodide.

Progress Check

1. What is an aquifer?
2. During purification, why is water filtered?
3. Why is water chlorinated?
4. Why is distillation not widely used to purify water?
5. What compounds found in fertilisers can cause environmental problems?

Exam Question

We can test for the presence of chloride ions in a solution by adding silver nitrate solution.

a. What would you see if the solution contained chloride ions?

b. What is the name of the silver compound that forms?

c. What is the state of the silver compound?

Hard and Soft Water

Water is a good solvent.

The Water Cycle

As the Sun warms the sea some of the water evaporates to form **water vapour**. This condenses to form clouds.

The water droplets in the clouds join together to produce rain, which falls back to the Earth's surface. In this way, water is continually moved around the water cycle.

Solubility

The **solubility** of a substance (solute) is measured as the number of grams of a substance that will dissolve in 100 g of water at a given temperature.

For solid substances, solubility generally increases as the temperature increases.

If a solution is **saturated**, then it cannot dissolve any more solute at that temperature. If the temperature of the solution is lowered, then some of the solute, that had been dissolved will come out of solution.

Gases

Many gases are soluble in water. The solubility of gases increases as the pressure increases and decreases as the temperature increases.

Carbonated drinks are made by dissolving **carbon dioxide** gas in water under high pressure. When the pressure is released, the gas comes out of solution and we see bubbles. A cold fizzy drink will contain more dissolved carbon dioxide than a warm fizzy drink.

The **oxygen** dissolved in water is essential for fish and other aquatic life. If the temperature of the water is increased, less oxygen is dissolved and aquatic life may perish.

Soft Water

Soft water does not contain calcium or magnesium ions. It readily forms a **lather** (bubbles) with soap.

Hard Water

Hard water contains calcium and magnesium ions. There are two types of hard water.

- Permanently hard water – this is caused by dissolved calcium sulfate and cannot be removed by boiling.
- Temporarily hard water – this is caused by dissolved calcium hydrogen carbonate and can be removed by boiling.

Water becomes hard when it comes into contact with rocks that contain calcium and magnesium ions. Hard water is good for you. It produces stronger bones and teeth and reduces the risk of developing heart disease.

Problems with Hard Water

Hard water, however, can also cause problems. More soap is required to form a lather with hard water than is required with soft water. In addition, **limescale** can form when hard water is heated and this can damage heating systems and kettles. Limescale can be removed by weak acids.

Hard water can be made into soft water by removing the calcium and magnesium ions. This can be done by adding **washing soda** (sodium carbonate). This reacts with the calcium or magnesium ions to form a precipitate of calcium carbonate or magnesium carbonate.

Water can also be softened by passing it through an **ion exchange column**. In the column, the calcium and magnesium ions are replaced by hydrogen and sodium ions.

Both hard and soft water form a good lather with soapless detergents.

1. In the water cycle, why does water evaporate?
2. Name the gas that is dissolved in water to make fizzy drinks.
3. For solids, how does increasing the temperature affect solubility?
4. For gases, how does increasing the temperature affect solubility?
5. For gases, how does increasing the pressure affect solubility?

EXAM QUESTION

Fizzy drinks contain dissolved carbon dioxide.

How does changing the temperature of the fizzy drink affect the amount of carbon dioxide that is dissolved in it?

Detergents

Detergents dissolve fat and grease. They are widely used as cleaning agents.

Washing Powders

We use washing powders to help get our clothes clean. They have lots of ingredients, including:

- **detergents** to remove the dirt from the fabric
- water softeners to remove the hardness from the water
- bleaches to remove stains
- **enzymes** that help to remove stains at low temperatures; this helps consumers to save money and helps to protect the environment
- optical brighteners to make the clothes look really clean.

Washing-up Liquids

We use washing-up liquids to help us to clean crockery. They have lots of ingredients, including:

- a detergent to get the crockery clean
- water to dilute the detergent so it is easier to use
- fragrance and colouring to make the washing-up liquid more attractive
- a water softener to make hard water softer
- a rinsing agent, which helps the water to drain away from the crockery.

Water is a good **solvent** for most ionic compounds. If a compound **dissolves**, we say it is **soluble** and the mixture that is made is called a solution.

Grease and fats do not dissolve well in water. We can use detergents to remove them from surfaces.

Many detergents are salts made by a neutralisation reaction between an acid and an alkali.

Detergent molecules have two parts: a **hydrophilic** head group that is attracted to water molecules and a **hydrophobic** part that avoids water but is attracted to fat or grease. The detergent molecules surround the fat or grease stain which can then be washed off.

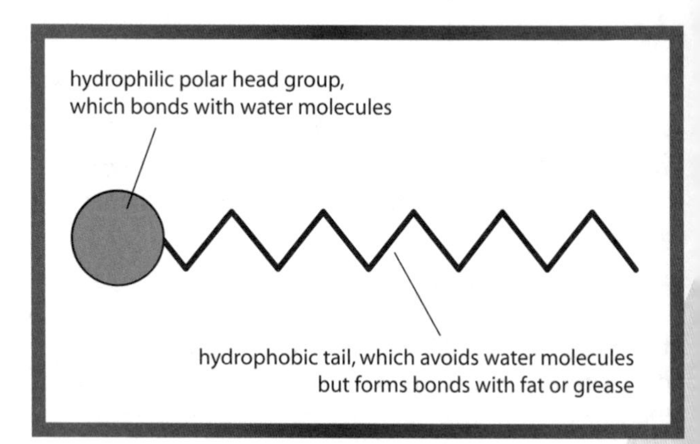

Washing Symbols

Washing symbols on clothes show the conditions that should be used to clean the garment.

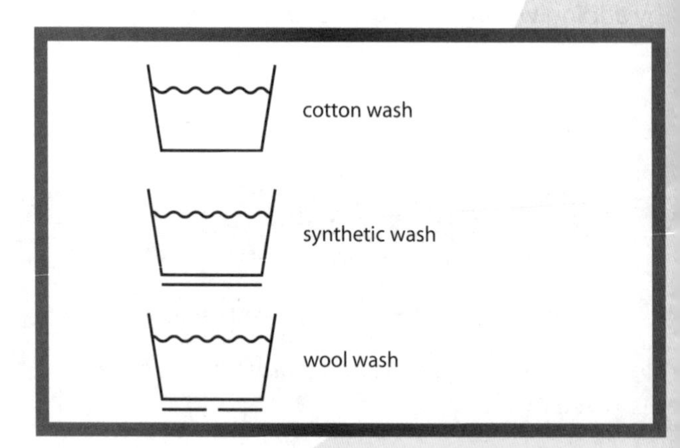

The temperature shown on the label is the maximum temperature that should be used. A lower temperature could be used but may not get the clothes as clean.

Dry Cleaning

Some fabrics can be damaged by washing them in water or may be stained with something that does not dissolve in water. These clothes can be dry cleaned.

In dry cleaning, the clothes are washed in a solvent other than water. This solvent is then extracted and the clothes are carefully pressed.

Biological and Non-Biological Detergents

Biological detergents contain enzymes.
Non-biological detergents do not contain enzymes.

1. Why are bleaches added to washing powders?
2. Why are enzymes used in some washing powders?
3. Why is colouring and fragrance added to washing-up liquids?
4. Why is a rinsing agent added to washing-up liquids?
5. On a washing symbol what does the number indicate?

EXAM QUESTION

Detergents are an important ingredient in washing powders.

a. Why are detergents used?

b. Many detergents are salts. What should be added to an alkali to make a salt?

c. Name the end of the detergent molecule that is attracted to water molecules.

Special Materials

Some materials have special properties that make them unique and extremely useful.

Smart Materials

Smart materials are very special materials that have one or more **property** that can be dramatically changed by changes in the environment.

Thermochromic pigments can be added to paints. If these paints are heated, the colour of the paint changes. Applications include designs on cups that appear when hot liquids are added.

Nitinol is another smart material. When a force is applied, nitinol stretches but when it is warmed up, it returns to its original shape.

Nano Particles

Nanoparticles consist of just a few hundred atoms, so they are incredibly small. Present uses of these materials include sunscreens. Future uses could include better, smaller computers.

Forms of Carbon

Diamond, graphite and fullerene are all **allotropes** (forms) of the element carbon.

Buckminsterfullerene, C_{60}

Fullerenes consist of cages of carbon atoms held together by strong covalent bonds. These cages can be used to hold other molecules. Scientists are developing ways of delivering drugs using fullerenes. The most stable fullerene is called buckminsterfullerene. It has the formula C_{60}.

Buckminsterfullerene is a black solid.

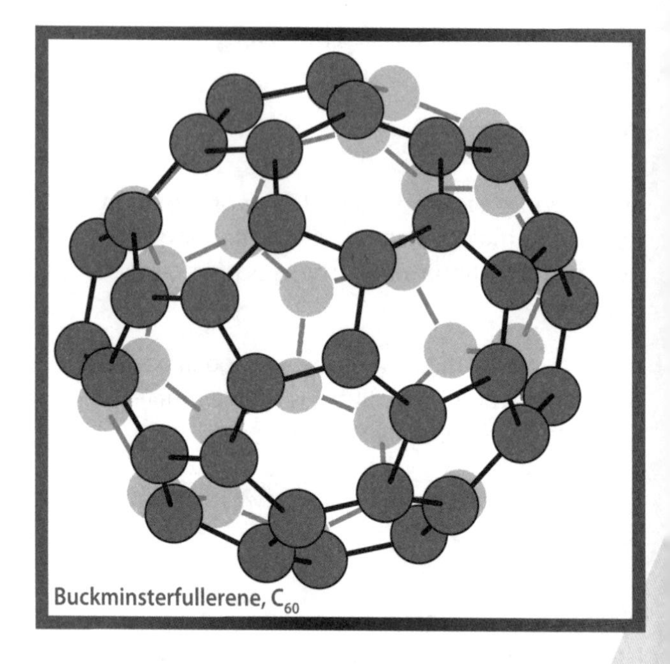

Buckminsterfullerene, C_{60}

Nanotubes

Fullerenes can be made into **nanotubes.** Nanotubes are very hard and strong. They can be made into lightweight sports equipment like tennis racquets. Other uses of nanotubes include industrial catalysts and semi-conductors for use in electrical circuits.

Diamond

Diamond is a highly valued gemstone. High-quality diamonds are **lustrous** and colourless and are used in jewellery. Diamonds can also be used to make cutting tools.

Diamond has a giant covalent structure. Each carbon atom is bonded to four other carbon atoms by strong covalent bonds. Diamond has a very high melting point and is hard because it has lots of bonds. It does not conduct electricity because there are no free electrons or ions to move.

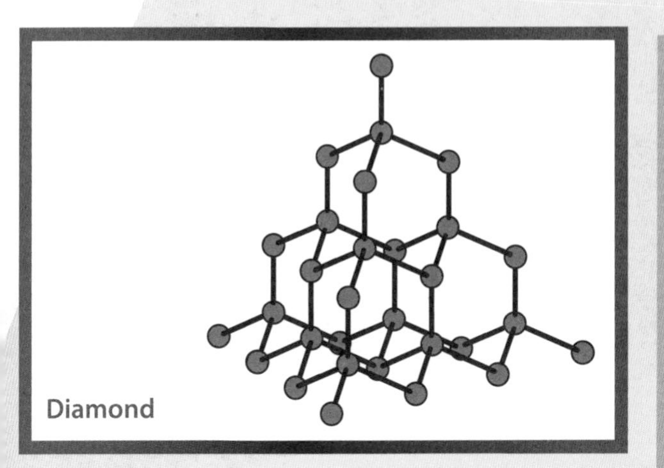

Diamond

Graphite

Graphite is used in pencil leads, as a lubricant and to make electrodes. In graphite, each carbon atom is bonded to three other carbon atoms in the same layer by strong covalent bonds.

Graphite has a high melting point because it has lots of strong covalent bonds. There are much weaker forces of attraction between the layers. This means that the layers of carbon atoms can pass easily over each other.

Graphite conducts electricity because the electrons in the weaker bonds between layers are able to move.

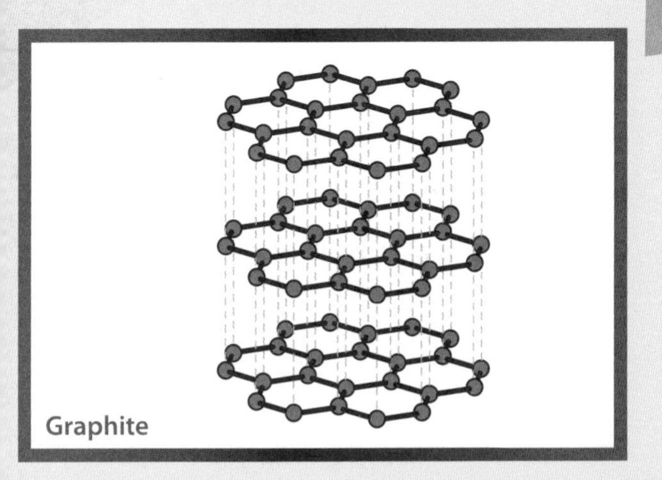

Graphite

1. Give **one** use of graphite.
2. What is the formula of buckminsterfullerene?
3. Name **three** allotropes of carbon.
4. Why are diamonds used in jewellery?
5. What colour is solid buckminsterfullerene?

Diamonds have a very high melting point and do not conduct electricity.

a. Describe the structure of diamond.

b. Why does diamond have a high melting point?

c. Why does diamond not conduct electricity?

New Materials

Many scientists work to produce useful new materials which help to improve people's lives

Gore-tex™

Gore-tex™ is used to make waterproof objects like jackets. It consists of a thin membrane, which is used to coat fabrics. The membrane has lots of little holes. Liquid water is too big to go through these holes, so the fabric is waterproof. Water vapour is small enough to pass through the holes so it is **breathable**.

In the past, waterproof jackets were made from nylon. This material was lightweight, hard wearing and waterproof but because it was not breathable, perspiration could make a nylon coat quite uncomfortable to wear.

Gore-tex™ jacket

PTFE

PTFE is better known by its trade name of 'Teflon'. Its properties are as follows:

- It is very unreactive.
- It has a very slippery surface.

PTFE was first discovered accidentally by scientists investigating refrigerant gases. Today it is used to make non-stick saucepans.

Kevlar

Kevlar is:

- lightweight
- very strong
- flexible.

Kevlar is used to make bulletproof vests.

Lycra

Lycra is a very **stretchy** material. It can be mixed with other fibres to make fabrics that can be used to make swimsuits.

Thinsulate

Thinsulate is an insulating material. It contains very small **fibres.** These fibres trap air which acts as a layer of insulation stopping body heat from escaping, so it keeps you warm.

Carbon Fibres

Carbon fibres:

- have a low density
- do not stretch
- are not compressed.

Carbon fibres are mixed with epoxy to make sports equipment such as squash racquets.

Phosphorescent Pigments

These pigments absorb light energy and then release it over a period of time. They can be used to make objects that glow in the dark such as wristwatch dials.

Analgesics

Drugs are substances that affect chemical reactions inside the body. Aspirin, paracetamol and ibuprofen are **analgesics**. These drugs are called painkillers.

All of these drugs have rings of carbon atoms.

Aspirin is made from salicylic acid. It was first discovered in willow bark but is now manufactured synthetically.

The benefits of aspirin include:

- it reduces pain
- it lowers body temperature
- it thins the blood to reduce the risk of blood clots.

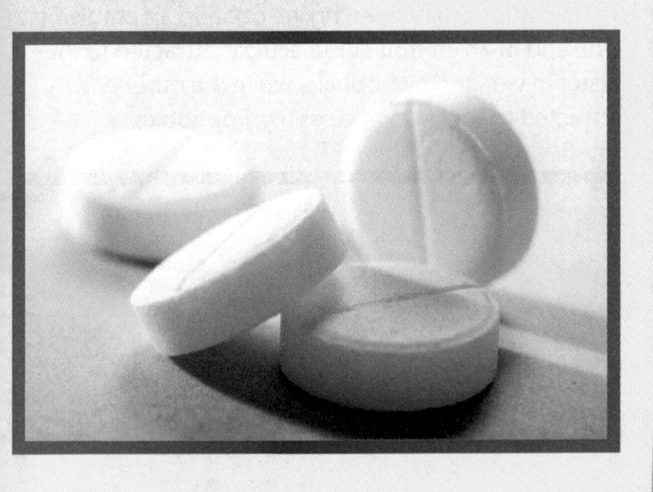

PROGRESS CHECK

1. Why can liquid water not pass through Gore-tex™?
2. Which material was widely used to make waterproof jackets in the past?
3. What is the trade name for PTFE?
4. How is PTFE used?
5. What is Kevlar used to make?

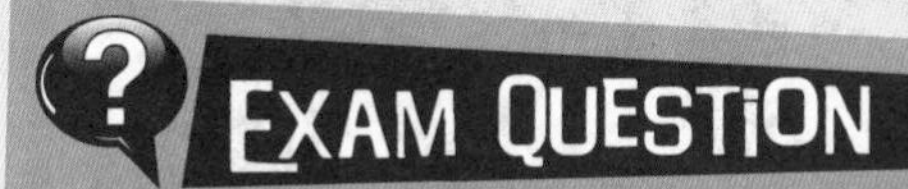

Thinsulate is a very useful material.

a. Describe the structure of thinsulate.

b. Explain why thinsulate might be used to make a hat.

Vegetable Oils

Vegetable oils are important foods. They are a good source of **energy** and of the vitamins A and D.

Oils as Foods

Popular vegetable oils include sunflower oil and olive oil.

Vegetable oils can also be used as **biofuels**. When they are burned, they release lots of energy.

Vegetable oils can be produced from:

- **seeds**
- **nuts**
- **fruits.**

First the plant material is crushed, then the oil is extracted.

Vegetable oils can be extracted from the fruits, seeds and nuts of some plants. The oil is removed by crushing up the plant material and collecting the oil

Vegetable oils are unsaturated because they contain carbon double bonds. We can check this by adding bromine water. The bromine water changes colour from brown to colourless.

Some vegetable oils contain many carbon double bonds. These are described as **polyunsaturated** fats. Doctors believe that polyunsaturated fats are good for our health.

Emulsions

An **emulsion** is a mixture of two immiscible (do not blend together) liquids.

Salad dressing is an example of an everyday emulsion. Salad dressings are made by shaking oil with vinegar so that the two liquids mix together, but the oil and vinegar soon separate out.

To stop this from happening we can add an **emulsifier**. One end of an emulsifier molecule is attracted to the water in vinegar (hydrophilic) while the other end is attracted to the oil molecules (hydrophobic).

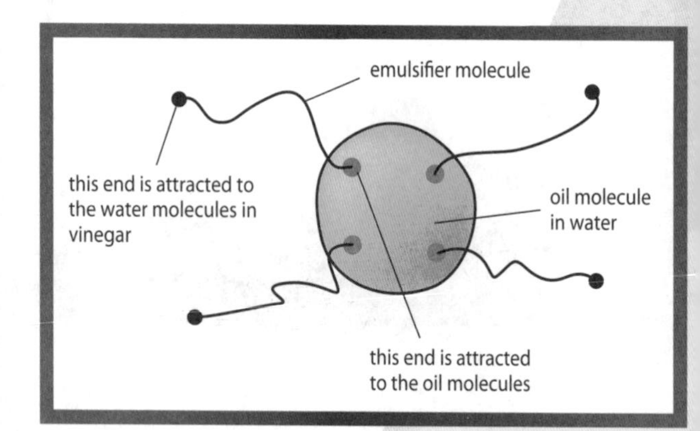

Milk is an oil in water emulsion.

Butter is a water in oil emulsion.

Hydrogenated Vegetable Oils

Most vegetable oils are liquids at room temperature. Liquid oils can be very useful, but there are times when we might prefer a solid fat, for example, when we want to spread it onto bread or make cakes.

Vegetable oils can be made solid at room temperature by a process known as hydrogenation. The vegetable oils are heated with hydrogen using a nickel catalyst at around 60°C.

Vegetable oils react with sodium hydroxide to form soaps.

fat + sodium hydroxide → soap + glycerol

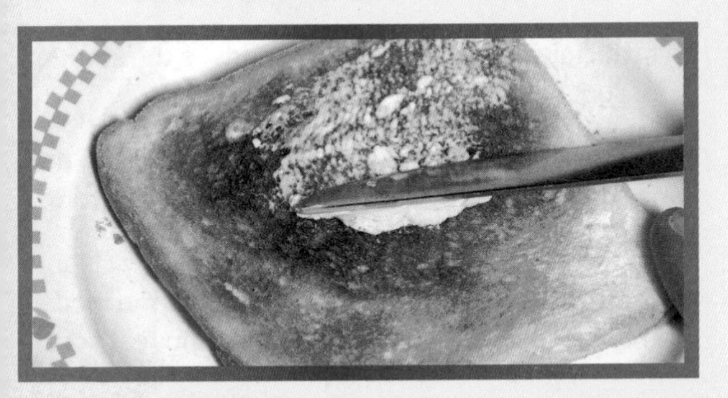

Paint

Paints can be used to make things look more attractive, or to protect them. Paint is a special type of mixture called a **colloid**. It consists of tiny coloured particles that are suspended in a solvent. Types of paint include oil paints and emulsion paints.

1. What vitamins are found in vegetable oils?
2. Name **two** types of vegetable oil.
3. Which parts of plants can we extract vegetable oils from?
4. Why can vegetable oils be described as 'unsaturated'?
5. What do salad dressings contain?

EXAM QUESTION

1. Use the words below to complete the table:

emulsion bromine water biofuel polyunsaturated

Name	Description
a.	A mixture of two immiscible liquids
b.	A fat with many carbon double bonds
c.	A material made from living organisms that can be burned to release energy
d.	A chemical used to test for a carbon double bond

Food Additives

Many foods are cooked before we eat them. The cooking of food is a chemical change.

Eggs and meat are good sources of **protein**. Potatoes are a good source of **carbohydrate**.

When we cook eggs or meat, the shape of the protein molecules are changed.

When we cook potatoes we make them easier to digest.

We cook food to make it:

- taste better
- look better
- easier to digest.

The high temperatures used in cooking can also kill micro-organisms in the food.

We sometimes add chemicals to foods to make it:

- look more attractive
- taste better
- have a longer shelf-life.

Chemicals that have passed safety tests and are approved for use throughout the European Union, are called **E-numbers**.

Food additives can be natural substances or artificial substances.

Chemicals commonly added to foods include:

- **emulsifiers** – used to keep unblendable liquids mixed together
- **colours** – used to make food look more attractive (artificial colourings can be detected by chromatography)
- **flavours** – used to make the food taste better
- **artificial sweeteners** – used to reduce the amount of sugar needed
- **preservatives** – these increase the shelf-life of the food by stopping harmful micro-organisms from growing
- **antioxidants** – increase the shelf-life of the foods that contain oils and fats by preventing reactions with oxygen.

Public Concern

The number of additives in food has caused some public concern. Scientists believe, however, that provided we eat a balanced diet including lots of different foods, the permitted food additives are safe for the vast majority of people.

Baking Powder

Baking powder is used to make biscuits and cakes. It contains **sodium hydrogen carbonate**, $NaHCO_3$.

When sodium hydrogen carbonate is heated, a thermal decomposition reaction takes place producing sodium carbonate, water and carbon dioxide.

sodium hydrogen carbonate	→	sodium carbonate	+	water	+	carbon dioxide
$2NaHCO_3$	→	Na_2CO_3	+	H_2O	+	CO_2

The products include carbon dioxide gas, which becomes trapped in the dough mixture and makes the dough rise. This gives cakes and biscuits a light, pleasing texture.

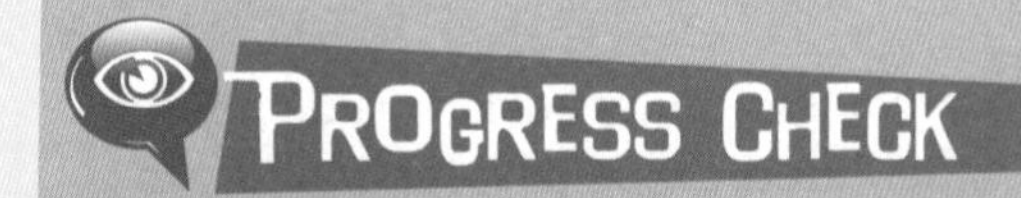

1. Where are E-numbers approved for use?
2. What does an emulsifier do?
3. Is a strawberry an artificial or a natural substance?
4. What type of additive can be added to a food to make it look better?
5. What type of food additive is added to oils and fats to stop them from reacting with oxygen?

Exam Question

1. What is the name of the chemical found in baking powder?
2. Why is baking power used when we make cakes or biscuits?
3. Complete the word equation to show the products of the thermal decomposition of sodium hydrogen carbonate:

   sodium hydrogen carbonate → sodium carbonate + water + __________

Crude Oil

Crude oil is a mixture. The most important compounds in crude oil are called **hydrocarbons**.

Fractional Distillation

Hydrocarbons are compounds that only contain carbon and hydrogen atoms.

Short hydrocarbon molecules are valuable **fuels** as they:

- are easy to ignite
- have low boiling points
- are runny.

Longer hydrocarbon molecules are less valuable, but they are still widely used. Crude oil is separated into **fractions** (groups of molecules with a similar number of carbon atoms) by fractional distillation.

During **fractional distillation** crude oil is heated up until it evaporates. Short hydrocarbon molecules have low boiling points and reach the top of the fractionating column before they condense and are collected.

Longer hydrocarbon molecules have higher boiling points; they condense and are collected lower down the fractionating column.

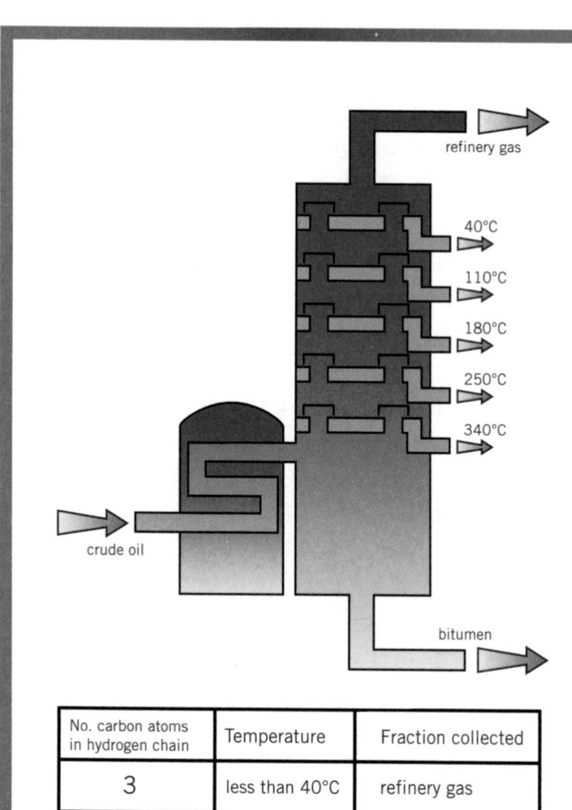

No. carbon atoms in hydrogen chain	Temperature	Fraction collected
3	less than 40°C	refinery gas
8	40°C	petrol
10	110°C	naphtha
15	180°C	kerosene
20	250°C	diesel
35	340°C	oil
50+	above 340°C	bitumen

A fractionating column

Uses

Fraction	Use
gases	heating
petrol	fuel for vehicles
naphtha	to make new chemicals
kerosene	jet fuel
diesel and oil	heating and as fuel for vehicles
bitumen	to make roads

Cracking

Fractional distillation of crude oil produces large amounts of long hydrocarbon molecules.

These long molecules can be broken down into smaller, more useful and more valuable molecules by **cracking**. In this process, large molecules are heated until they evaporate and are then passed over a catalyst.

Here decane is cracked to produce octane and ethene.

decane $C_{10}H_{22}$ (from the naphtha fraction) → octane C_8H_{18} + ethene C_2H_4

```
  H H H H H H H H H H            H H H H H H H H          H     H
  | | | | | | | | | |            | | | | | | | |           \   /
H-C-C-C-C-C-C-C-C-C-C-H   →   H-C-C-C-C-C-C-C-C-H   +      C=C
  | | | | | | | | | |            | | | | | | | |           /   \
  H H H H H H H H H H            H H H H H H H H          H     H
```

Cracking is an example of a thermal decomposition reaction

Progress Check

1. What is a hydrocarbon?
2. What is a fraction?
3. In a fractionating column, where are the shortest hydrocarbon molecules collected?
4. What is the name of the fraction which has the longest hydrocarbon molecules?
5. Which fraction is used for jet fuel?

Exam Question

Use the words below to complete the sentences.

condense **evaporates**
fractional distillation **fractions**

Crude oil is separated by ___a.___. The crude oil is heated until it ___b.___. Short hydrocarbon molecules reach the top of the column before they ___c.___ and are collected. Groups of molecules with a similar number of carbon atoms are called ___d.___.

Fuels

A fuel is a substance which can be burned to release **energy**. This means the burning of a fuel is an exothermic reaction.

A good fuel should:

- be easy to ignite
- be widely available
- release lots of energy when it is burned
- burn cleanly without producing lots of soot
- not damage the environment.

Most fuels contain carbon and/or hydrogen.

Hydrocarbons are compounds that only contain carbon and hydrogen atoms. Petrol, diesel and fuel oil are all hydrocarbons. Coal is mainly made of carbon.

Combustion

Complete combustion of hydrocarbon fuels produces **carbon dioxide, CO_2** and **water vapour, H_2O.**

Scientists believe that carbon dioxide contributes towards global warming.

Incomplete combustion of hydrocarbon fuels can occur if there is a limited supply of oxygen. This can occur in faulty gas appliances. The products of incomplete combustion include **carbon monoxide, CO** and carbon.

Carbon monoxide is a toxic gas. It lowers the ability of red blood cells to carry oxygen around the body.

Unburnt carbon makes a flame look yellow and it is deposited as **soot**, making surfaces dirty.

Hydrogen

Hydrogen is used as a fuel in space rockets, but currently it is not used as a fuel for cars.

Advantages of using hydrogen as a fuel	Disadvantages of using hydrogen as a fuel
Releases lots of energy when it is burned	It is a gas that can only be liquefied at very low temperatures or very high pressures
Burns very cleanly	It is difficult to store – a car would require a large, heavy container to store the hydrogen
The only product of combustion is water vapour which does not harm the environment	Hydrogen is not found naturally and would have to be made before it could be used. This would probably require the use of fossil fuels

Biofuels

Biofuels are fuels that are produced from living things. Wood and ethanol are examples of biofuels.

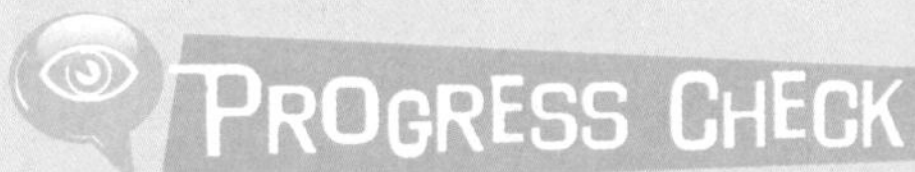

PROGRESS CHECK

1. Which **two** elements are found in hydrocarbon fuels?
2. What are the products of the complete combustion of hydrocarbon fuels?
3. Name the main element found in coal.
4. What is the environmental problem associated with carbon dioxide?
5. Why is the incomplete combustion of hydrocarbon fuels undesirable?

EXAM QUESTION

1. Complete the word equation below to show the product of the complete combustion of carbon.

   Carbon + oxygen → ____________

2. Carbon can also be burned in a limited supply of oxygen. Name one of the products of the incomplete combustion of carbon.
3. Many fuels also contain atoms of hydrogen. Complete the word equation to show the product of the combustion of hydrogen.

   hydrogen + oxygen → ____________

Air and Air Pollution

Today's atmosphere is composed of about 78% nitrogen, about 21% oxygen, about 1% argon and small amounts of other gases including carbon dioxide and water vapour. We can obtain nitrogen and oxygen from liquid air by fractional distillation.

Evolution of the Atmosphere

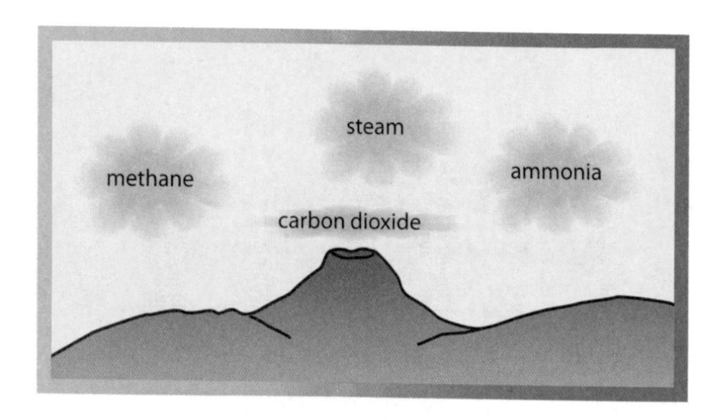

During the first billion years of the Earth's life, there were enormous amounts of volcanic activity. The volcanoes released water vapour, which condensed to form the early **oceans**, while the other gases that were released formed the Earth's early atmosphere.

This atmosphere mainly consisted of **carbon dioxide**, like the present-day atmosphere of Mars and Venus.

During the next two billion years, **plants** evolved and began to cover the Earth's surface. These plants steadily removed carbon dioxide and produced oxygen by photosynthesis.

In recent times, people have started to burn large amounts of fossil fuels. This has released large amounts of carbon dioxide back into the atmosphere.

Acid Rain

Many fossil fuels contain small amounts of **sulfur**. When these fuels are burned, this sulfur reacts with oxygen to form sulfur dioxide. If this gas is released into the atmosphere, it can react with rainwater to form acid rain.

We can reduce the amount of acid rain produced by:

- burning less fossil fuel
- removing sulfur compounds from oil and natural gas
- removing sulfur dioxide from the waste gases in coal-powered power stations before they are released into the atmosphere.

Global Dimming

Global dimming is caused by smoke particles released into the atmosphere when fuels are burned.

Global Warming

Scientists have found that the temperature of the Earth seems to be gradually increasing.

Carbon dioxide, which is produced when fossil fuels are burned, is thought to be contributing to this effect known as global warming.

The carbon dioxide traps heat energy reaching the Earth from the Sun.

If global warming continues, the polar icecaps may eventually melt. This could cause massive environmental problems.

Nitrogen Oxides

At the high temperatures inside car engines, nitrogen in the atmosphere may react with oxygen to form **nitrogen monoxide**. The nitrogen monoxide is then oxidised to form **nitrogen dioxide**. Nitrogen monoxide and nitrogen dioxide are known as nitrogen oxides or NO_x.

Reducing Pollution from Cars

The amount of atmospheric pollution caused by cars can be reduced by:

- using more efficient engines
- using public transport more
- using catalytic converters; these convert carbon monoxide to carbon dioxide and nitrogen monoxide to nitrogen and oxygen.

1. What is the main gas in today's atmosphere?
2. What was the main gas in the Earth's early atmosphere?
3. Why did the amount of carbon dioxide in the Earth's early atmosphere decrease?
4. Name the gas which makes up around 20% of today's atmosphere.
5. Name the gas produced when sulfur reacts with oxygen.

EXAM QUESTION

The Earth's early atmosphere mainly consisted of carbon dioxide.

a. Which planet has an atmosphere similar to the Earth's early atmosphere?

b. Explain why the Earth's atmosphere today has less carbon dioxide and more oxygen than it has had in the past.

c. What is the environmental problem associated with increased levels of carbon dioxide in the atmosphere?

Pollution

Limestone is an important raw material. It can be used as a building material and can be made into other important materials like cement and glass.

Limestone

The advantages of limestone quarrying include the economic and social benefits of new jobs for local people.

The disadvantages of a limestone quarry include noise and dust from the quarry and problems created by lorries transporting the limestone.

Plastics

Plastics are a type of **polymer**. The properties of plastics depend on how they are made and what they are made from.

Polymers are very widely used and new applications are being developed including new **'intelligent' packaging.** 'Intelligent' packaging is used to improve the quality and safety of foods; for example, they can be used to remove water from inside a packet so that it is more difficult for bacteria or mould to grow and the food will stay fresh for longer.

Problems with Plastics

Plastics are very useful. Most plastics, however, are **non-biodegradable**. This means that when plastic objects are thrown away, they remain in the environment. This can cause problems:

- Landfill sites fill up more quickly.
- If we try to get rid of the plastics by burning them, toxic gases may be produced.
- It is very difficult to recycle plastics because it is hard to separate them out.

Because of these problems, scientists have developed a range of plastics that are biodegradable.

Recycling

Recycling waste materials, including glass, metal and paper, helps to protect the environment because fewer raw materials are required. This also means that less waste needs to be disposed of.

Sustainable Development

Sustainable development balances the need for **economic development** with a respect for the **environment**, so that people can enjoy a good standard of living today without compromising the needs of future generations.

Progress Check

1. How can a new limestone quarry benefit people?
2. What are the disadvantages of having a limestone quarry nearby?
3. Name a type of polymer.
4. What could be a disadvantage of burning plastics?
5. Why is it difficult to recycle plastics?

Exam Question

Scientists have developed 'intelligent' packaging for food.

a. What can 'intelligent' packaging do?

b. Why is this an advantage to consumers?

Alkanes and Alkenes

Carbon atoms form four bonds with other atoms while hydrogen atoms only form one bond.

Alkanes

The **alkanes** are a family of **hydrocarbon** molecules. Hydrocarbon molecules only contain carbon and hydrogen atoms. Alkanes are saturated hydrocarbons because they do not contain carbon double bonds. Short alkane molecules are useful fuels.

Alkanes have the general formula C_nH_{2n+2}

Name	methane	ethane	propane	butane
Chemical formula	CH_4	C_2H_6	C_3H_8	C_4H_{10}
Structure	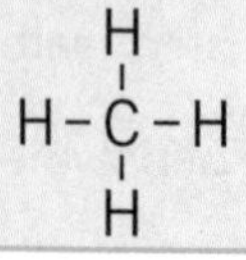 H–C–H with H above and below the C	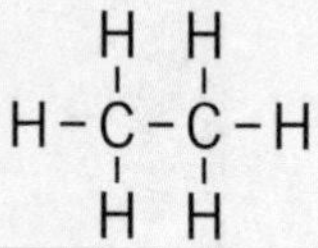 H–C–C–H with H above and below each C	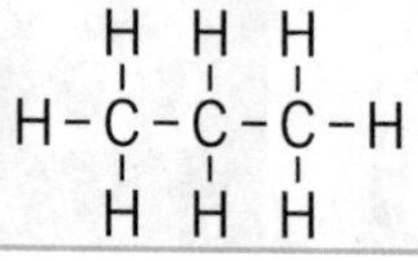 H–C–C–C–H with H above and below each C	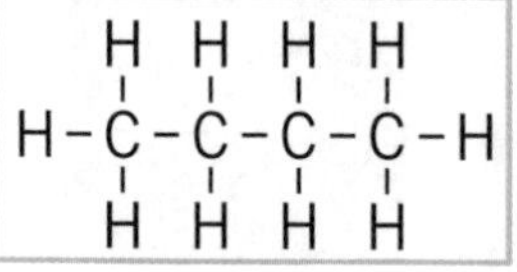 H–C–C–C–C–H with H above and below each C

The lines in these structural diagrams represent covalent bonds

Alkenes

The **alkenes** are another family of hydrocarbon molecules.

Alkenes are produced during cracking.

Alkenes are unsaturated hydrocarbons because they do contain carbon double bonds. Alkene molecules are more reactive than alkane molecules.

They can be used to make new chemicals including plastics.

Alkenes have the general formula C_nH_{2n}.

Name	ethene	propene
Chemical formula	C_2H_4	C_3H_6
Structure	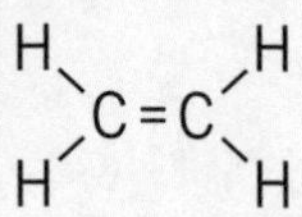 C=C with two H on each C	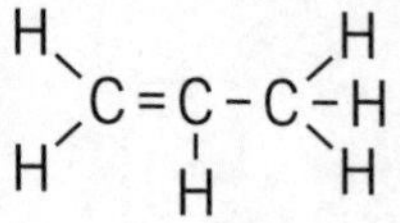 C=C–C: two H on the first C, one H on the second C, three H on the third C

Industrial Alcohol

Ethanol can be made from non-renewable sources. Ethanol produced in this way is called industrial alcohol. Ethene, which is produced by the cracking of long chain hydrocarbons, is reacted with steam to produce ethanol.

ethene + steam → ethanol

$C_2H_4 + H_2O \rightarrow C_2H_5OH$

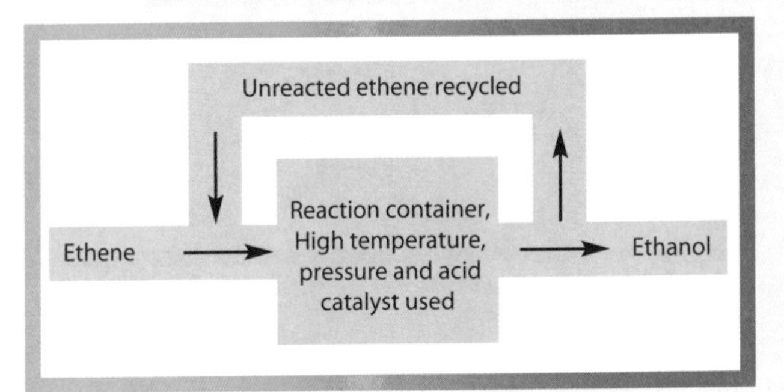

Uses of Alcohol

Ethanol can be used as a fuel. It burns very cleanly producing very little carbon monoxide. Ethanol can also be made by fermentation. During fermentation glucose reacts to produce ethanol and carbon dioxide.

The glucose can be obtained from sugar cane or sugar beet. Ethanol produced in this way is a renewable fuel.

Progress Check

1. What is the general formula of an alkane?
2. Give a use for a short alkane molecule.
3. What do the lines in structural diagrams represent?
4. How can alkenes be made?
5. What is the formula of ethene?

Exam Question

Ethanol can be made by reacting ethene with steam.

a. What family does ethene belong to?

b. Ethene is 'unsaturated' – what does this mean?

c. Write a word equation for the reaction of ethene with steam.

Alcohols, Acids and Esters

Alcohols are a family of organic compounds with the functional group hydroxyl, OH.

Alcohols

Short carbon chain alcohols are miscible with water (mix fully with water). The longer the carbon chain, the less soluble the alcohol.

Alcohols are neutral compounds (they have a pH of 7). They are widely used as solvents.

Methanol

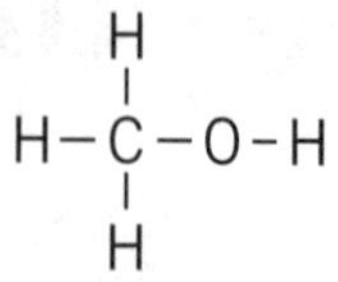

```
    H
    |
H - C - O - H
    |
    H
```

Ethanol

Ethanol (often just called alcohol) is also used as a fuel and is found in alcoholic drinks.

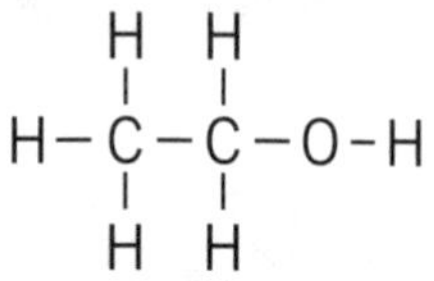

```
    H   H
    |   |
H - C - C - O - H
    |   |
    H   H
```

Carboxylic Acids

Carboxylic acids are another group of organic compounds. They have the carboxyl, COOH, functional group.

Examples include:

Methanoic Acid

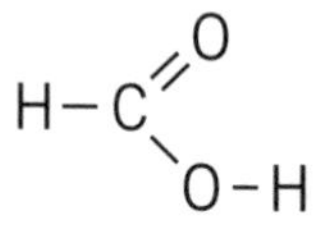

```
      O
     //
H - C
     \
      O - H
```

Ethanoic Acid

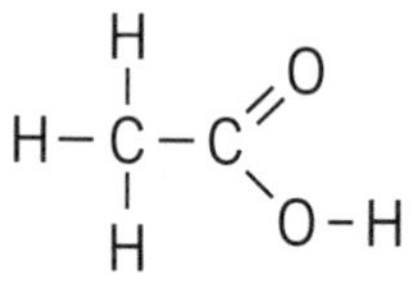

```
    H     O
    |    //
H - C - C
    |    \
    H     O - H
```

Carboxylic acids have a sharp, distinctive odour. They are all weak acids and react with bases to form salts. Solutions of carboxylic acids have a pH less than 7.

Esters

Esters can be made by reacting alcohols with carboxylic acids.

ethanol + ethanoic acid ⇌ ethyl ethanoate + water

$CH_3CH_2OH + CH_3COOH \rightleftharpoons CH_3COOCH_2CH_3 + H_2O$

Ethyl Ethanoate

```
                  O         H
      H   H        \\       |
      |   |          C - C - H
  H - C - C - O  /          |
      |   |                 H
      H   H
```

Esters have pleasant, sweet odours and in nature are responsible for the taste and smell of many fruits.

Synthetic esters are widely used as flavourings in chocolates, sweets and other processed foods. They are also used in cosmetics like body sprays and as solvents in perfumes and aftershaves.

1. What family does ethanol belong to?
2. What family does ethyl ethanoate belong to?
3. What family does methanol belong to?
4. What family does methanoic acid belong to?
5. Suggest a pH for ethanoic acid.

EXAM QUESTION

Ethanol is a type of alcohol. It has the molecular formula CH_3CH_2OH.

a. A sample of ethanol is tested using a pH meter. What do you expect the pH to be?

b. Give a use of ethanol.

Polymers

In **addition polymerisation** many small molecules are joined together to form a bigger molecule.

Polymerisation

The small molecules are called **monomers.** These monomers are unsaturated molecules. The bigger molecule is called a polymer. Plastics are **polymers**.

This diagram represents the reaction between lots of ethene molecules to form polythene.

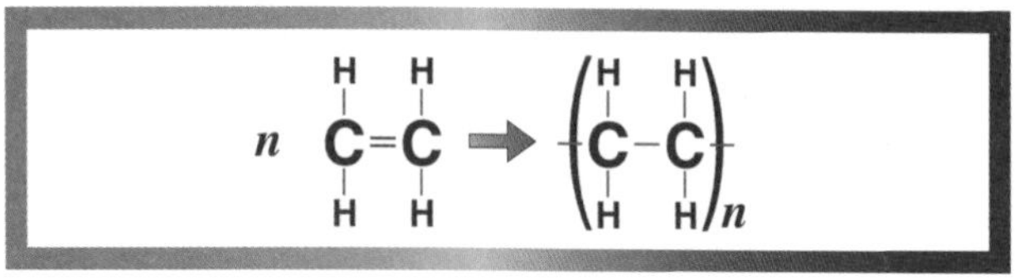

$$n\ \begin{matrix} H & & H \\ | & & | \\ C & = & C \\ | & & | \\ H & & H \end{matrix} \rightarrow \left(\begin{matrix} H & & H \\ | & & | \\ -C & - & C- \\ | & & | \\ H & & H \end{matrix} \right)_n$$

This diagram represents the reaction between lots of propene molecules to form polypropene.

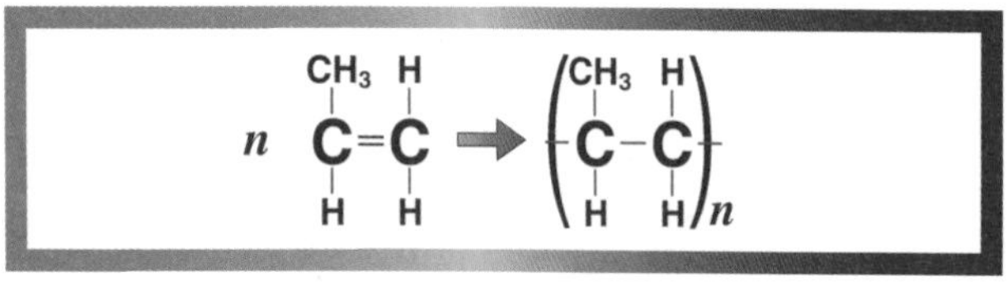

$$n\ \begin{matrix} CH_3 & & H \\ | & & | \\ C & = & C \\ | & & | \\ H & & H \end{matrix} \rightarrow \left(\begin{matrix} CH_3 & & H \\ | & & | \\ -C & - & C- \\ | & & | \\ H & & H \end{matrix} \right)_n$$

Properties of Plastics

The structure and bonding within a material affects its properties.

The stronger the forces between the particles in a solid, the higher the temperature at which the solid melts.

Modifying the structure of a polymer can influence its properties.

- Adding a **plasticiser** can make a plastic more flexible.
- Lengthening the polymer chain increases the forces of attraction between the molecules and increases the melting point of the polymer.
- Making the polymer chains more aligned increases the forces of attraction between the molecules and increases the melting point.
- Some plastics consist of long polymer chains with very little cross linking between the chains. When these plastics are heated, the chains untangle and the plastic softens. This means that these plastics can be reshaped many times.
- Other plastics consist of polymer chains that are heavily **cross linked**. These polymers must be shaped when they are first made. When they are heated again they will not soften but may eventually burn.

Plastics are useful but disposing of plastics can be difficult

Progress Check

1. Name the small molecules used to make polymers.
2. Name the polymer made from ethene.
3. Name the polymer made from propene.
4. How does adding a plasticiser affect a plastic?
5. What happens when a heavily cross-linked polymer is heated?

Exam Question

1. Name the monomer used to make polythene.
2. Which hydrocarbon family does the monomer used to make polythene come from?
3. What is the formula of an ethene molecule?

Limestone

Limestone is a type of sedimentary rock. It is mainly made of **calcium carbonate**, $CaCO_3$.

Heating Limestone

When limestone is heated it decomposes to form calcium oxide and carbon dioxide:

calcium carbonate → calcium oxide + carbon dioxide

This is an example of a **thermal decomposition** reaction.

Calcium oxide is also known as **quicklime**. Quicklime can be reacted with water to form calcium hydroxide:

calcium oxide + water → calcium hydroxide

Calcium hydroxide is also known as **slaked lime**.

Other Metal Carbonates

Other metal carbonates react in a similar way when they are heated.

More Useful Materials

Other useful materials made from limestone include:

- cement – made by heating powdered limestone and powdered clay and then adding water
- mortar – made by mixing cement, sand and water
- concrete – made by mixing cement, sand, rock chippings and water
- glass – made by heating up a mixture of limestone, sand and soda until it melts.

Reinforced concrete is a composite material made by allowing concrete to set around steel supports.

Clues in Names

The name calcium carbon**ate** tells us the compound contains calcium, carbon and lots of oxygen.

The names of some other compounds such as potassium nitr**ite** have a slightly different ending.

If a compound ends in 'ite', it contains some oxygen but not as much as if the name of the compound ended in 'ate'.

Potassium nitrite, therefore, must contain potassium, nitrogen and some oxygen.

Making Salts

Salts are very important compounds. They can be made by reacting acids with metal carbonates, metal oxides or metal hydroxides.

Sulfuric acid forms sulfate salts while hydrochloric acid forms chloride salts.

Other Rocks

- Granite is an example of an igneous rock. Igneous rocks are formed when molten rocks cool down and solidify.
- If the rock formed quickly, it will contain small crystals, for example basalt.
- If the rock formed slowly, it will contain large crystals, for example gabbro.
- If limestone is subjected to high temperatures or pressures, it can be turned into the metamorphic rock marble.

Progress Check

1. Name the main chemical compound in limestone.
2. What type of rock is limestone?
3. Name the products of the thermal decomposition of limestone.
4. By what name is calcium oxide also known?
5. By what name is calcium hydroxide also known?

Exam Question

1. What type of rock is granite?
2. Name the rock formed when limestone is subjected to high temperatures and pressure.

Structure of the Earth

The Earth has a layered structure.

The Layered Structure

- At the centre of the Earth is the **core**. The core is divided into two parts: the solid inner core and the outer core, which is liquid.
- The core is surrounded by the **mantle**.
- Around the mantle is the **crust**.

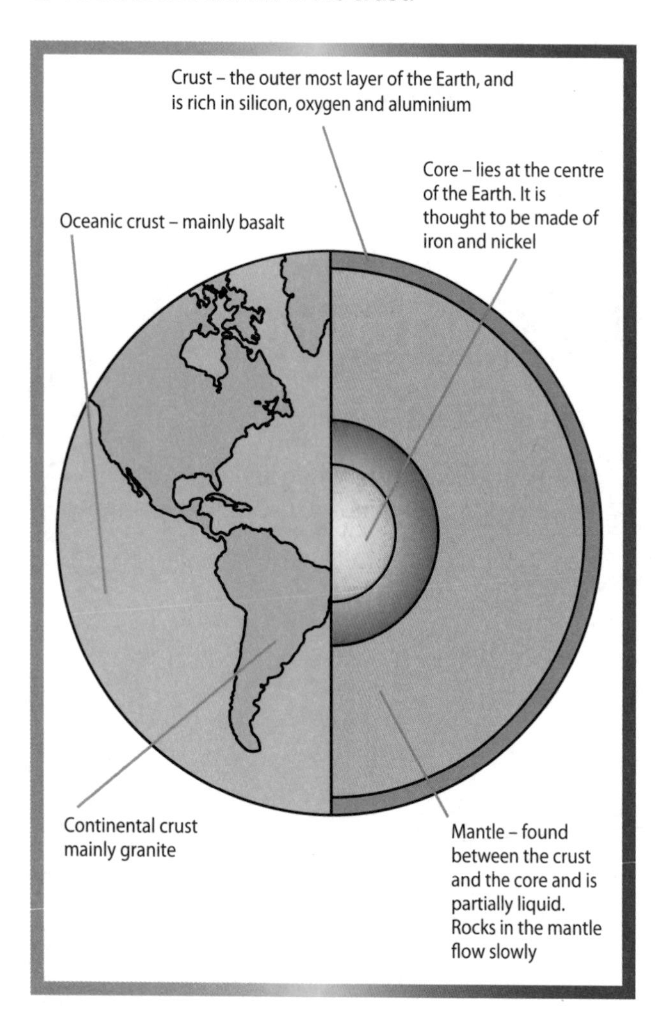

Plate Tectonics

People used to believe that the features that we see at the Earth's surface were formed as the Earth cooled down. Scientists now believe that these features were caused by **plate tectonics**.

In this theory, the Earth's **lithosphere** (crust and the upper part of the mantle) is broken into about a dozen pieces called plates. These plates are carried by convection currents in the Earth's mantle. The currents are caused by the heat released by natural radioactive decay. The plates move a few centimetres each year. This is about the same rate as your finger nails grow.

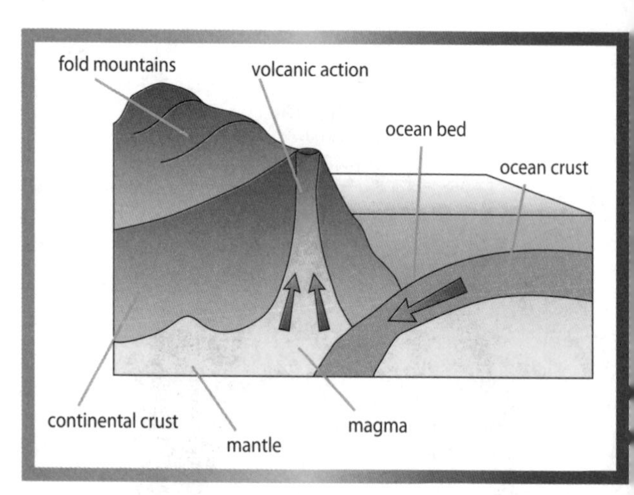

Earthquakes

Sometimes the plates cannot move smoothly. **Earthquakes** occur where the plates try to move past each other but become stuck. The forces on these plates gradually build up. Eventually the plates move and they release the strain that has built up as an earthquake, or if it happens under water, a **tsunami.**

Volcanoes

Volcanoes are also found at plate boundaries. When an oceanic plate collides with a continental plate, the denser oceanic plate is subducted beneath the continental plate. Some of the oceanic plate may melt to form magma (molten rock that is underground). The hot magma may be less dense than the surrounding rock; if it is, it may rise to the surface through cracks to form volcanoes. Volcanoes can cause devastating loss of life.

If the magma is iron-rich, it will be quite runny and the volcano will erupt relatively slowly and safely.

If the magma is silica-rich, however, it will be much more viscous. The volcano will erupt explosively, producing volcanic ash and throwing out molten rock called 'bombs'. These volcanoes are very dangerous to local people.

Although our methods of predicting when earthquakes and volcanoes will happen are improving, we still cannot say exactly when they will occur.

1. In what state is the inner core?
2. In what state is the outer core?
3. What is the name of the layer between the outer core and the crust?
4. Roughly how many plates are there?
5. What can form if an earthquake happens under the sea?

EXAM QUESTION

The Earth's lithosphere is split into about a dozen plates.

a. What is the lithosphere?

b. At what rate do these plates move?

c. Historically, how did people believe that mountains formed?

Cosmetics

A solution is a mixture made when a solvent dissolves a solute.

Nail Varnish

Water is a good solvent for many, but not all, substances. Some solutes like nail varnish are insoluble in water.

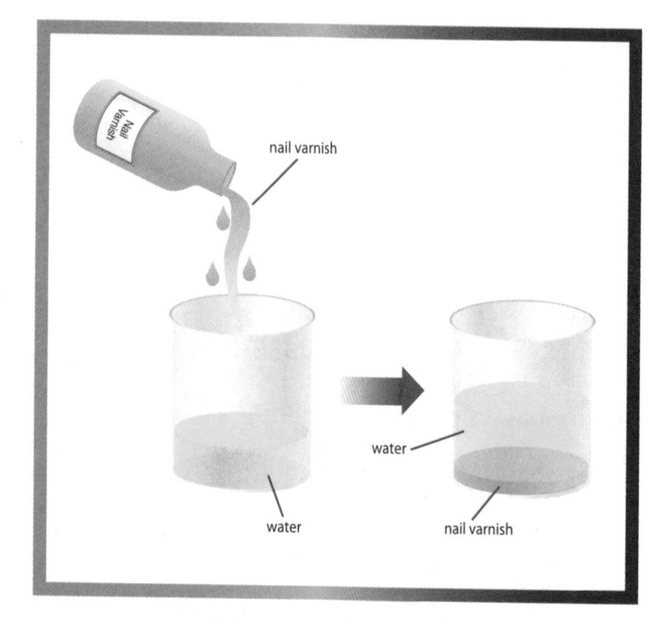

Perfumes

Ethanol is often used as a solvent in perfumes. Traditional perfumes contain plant and animal extracts such as jasmine and lavender, but these ingredients can be very expensive.

Today we often use cheaper, man-made fragrances such as esters. Esters are made by reacting carboxylic acids with alcohols.

A good perfume should:

- evaporate easily from the skin
- be non-toxic
- not react with water (sweat)
- not irritate the skin
- be insoluble in water so it is not washed off easily.

It is also important that perfume evaporates easily from the skin.

Perfumes evaporate because although there are strong forces of attraction within perfume molecules, there are much weaker forces of attraction between perfume molecules.

When the perfume is put on the skin, some of the molecules gain enough energy to evaporate.

Animal Testing

Cosmetic products have to be tested before they can be sold. Some tests involve living animals. Some people believe that this causes avoidable suffering to animals, while other people think that animal testing is the best way to ensure that products are safe for people to use.

Making Ethanol

Ethanol is a type of alcohol. It has the structure:

```
    H   H
    |   |
H - C - C - O - H
    |   |
    H   H
```

Ethanol can be made from sugar cane or sugar beet. It is a useful biofuel (a fuel made from living materials). Unlike petrol, ethanol is a renewable fuel but large areas of fertile land must be used to grow the sugar cane or sugar beet.

Fermentation

Ethanol is produced by **fermentation**. During fermentation, yeast converts glucose (sugar) into ethanol and carbon dioxide:

$$\text{glucose} \xrightarrow{\textit{yeast}} \text{ethanol} + \text{carbon dioxide}$$

$$C_6H_{12}O_6 \rightarrow 2C_2H_5OH + 2CO_2$$

Fermentation has been used to produce alcoholic drinks such as wine and beer. The consumption of alcoholic drinks causes many social and health problems and is banned by some religions.

1. What is a solution?
2. Name a solute which is insoluble in water.
3. Name a solvent often used in perfumes.
4. Write down a plant extract used in perfumes.
5. How can an ester be made?

EXAM QUESTION

Ethanol is a type of alcohol.

a. What is the name of the reaction in which glucose is converted into alcohol and carbon dioxide?

b. Write a word equation for this reaction.

Chemicals

It is very useful to be able to identify chemicals.

Uses

Different chemicals have different uses.

Chemical	Use
Ammonia	To make fertilisers
Carbohydrate	A type of food we need for energy
Carbon dioxide	To make carbonated drinks
Caustic soda	To make bleach
Citric acid	To flavour foods and drinks
Ethanoic (acetic) acid	In vinegar
Hydrochloric acid	To remove limescale from boilers
Phosphoric acid	A catalyst for the production of industrial alcohol
Sodium chloride	To 'grit' roads in winter
Water	In industry and in agriculture

Gas Tests

Hydrogen

Burns with a 'squeaky' pop.

Oxygen

Relights a glowing splint.

Carbon dioxide

When bubbled through limewater, it turns the limewater cloudy.

Ammonia

Turns damp red litmus paper blue.

Chlorine

Bleaches damp litmus paper.

Collecting Gases

- Upward delivery is used to collect gases that are less dense than air.
- Downward delivery is used to collect gases that are denser than air.
- Gases that are fairly insoluble are collected over water.
- If we need to measure the volume of gas, we can collect it using a gas syringe.

Hazard Symbols

Hazard symbols warn us about chemicals:

Oxidising
Allows other materials to burn more fiercely.

Highly flammable
Catches fire easily.

Toxic
Can cause death.

Harmful
Less dangerous than toxic.

Corrosive
Attacks and destroys living tissue.

Irritant
Can cause reddening or blistering of skin.

Flame Tests

Flame tests can be used to identify the metals in compounds.

The colour of the flame indicates the **metal** present.

- lithium – red
- potassium – lilac
- barium – apple green
- sodium – orange
- calcium – brick red

Hydroxide Tests

We can identify the metals present in metal salt solutions by adding **sodium hydroxide solution.**

If the metal ion forms a precipitate, we can use the colour of the **precipitate** to identify the metal present.

- copper(II) – pale blue precipitate
- iron(II) – green precipitate
- iron(III) – brown precipitate

Progress Check

1. Which gas burns with a 'squeaky' pop?
2. What type of chemical allows other chemicals to burn more easily?
3. What type of chemical can cause death?
4. What does a corrosive chemical do?
5. What colour would you expect when a sample of a sodium salt is used in a flame test?

Other Tests

Ion	What to do	What happens
Ammonium	React with sodium hydroxide then test gas produced (ammonia)	Damp red litmus goes blue
Nitrate	React with aluminium powder and sodium hydroxide then test the gas produced (ammonia)	Damp red litmus goes blue
Sulfite	React with dilute hydrochloric acid test the gas given off (sulfur dioxide)	Potassium dichromate(VI) solution turns from orange to green
Carbonate	React with dilute acid then bubble the gas which is produced through limewater (carbon dioxide)	Limewater goes cloudy

Modern instrumental methods are accurate, sensitive and rapid.

Exam Question

Use the words below to complete the sentences.

carbon dioxide **ammonia**
chlorine **oxygen**

If __a.__ is bubbled through limewater, the limewater turns cloudy. __b.__ relights a glowing splint.
__c.__ bleaches damp litmus paper. __d.__ turns damp red litmus paper blue.

Answers

Day 1

pages 4–5
How Science Works
PROGRESS CHECK
1. The variable we choose to change in an experiment
2. The variable that we measure in an experiment
3. A variable that can be put in order
4. A variable that can have any whole number value
5. It is close to the true value

EXAM QUESTION
a. The force applied
b. The length of the spring
c. Using the same spring, etc.

pages 6–7
Atomic Structure
PROGRESS CHECK
1. They have the same number of protons
2. Nucleus
3. Electrons
4. In the periodic table
5. Roughly 100

EXAM QUESTION
a. Calcium carbonate + hydrochloric acid → calcium chloride + water + carbon dioxide
b. One carbon atom and two oxygen atoms

pages 8–9
The Periodic Table
PROGRESS CHECK
1. Groups
2. Periods
3. Number of protons added to number of neutrons
4. Atomic number/proton number/number of protons
5. Mass number/number of neutrons

EXAM QUESTION
6 protons, 6 electrons and 7 neutrons

pages 10–11
Electronic Structure
PROGRESS CHECK
1. Shells/levels
2. Two
3. 2,1
4. Group 1
5. Group 4

EXAM QUESTION
a. 2,8,2
b. Group 2

pages 12–13
Ionic Bonding
PROGRESS CHECK
1. An atom or group of atoms with a charge
2. 1–
3. 1+
4. 2–
5. 2+

EXAM QUESTION
a. It has lots of strong ionic bonds
b. Sodium + chlorine → sodium chloride
c. 1+

pages 14–15
Covalent Bonding
PROGRESS CHECK
1. H_2
2. HCl
3. CH_4
4. NH_3
5. O_2

EXAM QUESTION
a. Shared pair of electrons
b. H_2O
c. Although there are strong bonds within the water molecules there are only weak forces of attraction between one molecule and the next

pages 16–17
Alkali Metals
PROGRESS CHECK
1. On the left-hand side
2. Potassium
3. Hydrogen
4. Orange
5. Lilac

EXAM QUESTION
a. Sodium + water → sodium hydroxide + hydrogen
b. $2Na + 2H_2O \rightarrow 2NaOH + H_2$
c. So it does not react with air or water
d. The outer electron is further from the nucleus so it is lost more easily

Day 2

pages 18–19
Noble Gases and Halogens
PROGRESS CHECK
1. Halogens/Group 7
2. Noble gases/Group 0
3. Liquid
4. Balloons
5. Filament lamps

EXAM QUESTION
a. Halogens/Group 7
b. Chlorine + potassium bromide → potassium chloride + bromine

pages 20–21
Calculations 1

PROGRESS CHECK

1. To compare the mass of different atoms
2. Carbon-12
3. The relative formula mass of a substance in grams
4. It does not go to completion
5. The percentage yield of a reaction = actual amount of product / theoretical yield × 100%

EXAM QUESTION

a. The reaction is reversible and does not go to completion/ some of the product was lost for example during filtering or evaporation/there may be side-reactions that are producing another product
b. 88%

pages 22–23
Calculations 2

PROGRESS CHECK

1. a. 0.05
   b. 0.2
   c. 0.25
2. a. 0.0025
   b. 0.01

EXAM QUESTION

Phenolphthalein

pages 24–25
Calculations 3

PROGRESS CHECK

1. The ratio of atoms in its simplest form
2. CH_2
3. NH_2
4. kJ mol^{-1}
5. Exothermic

EXAM QUESTION

The reaction is exothermic, overall energy is given out

pages 26–27
Haber Process

PROGRESS CHECK

1. Fractional distillation of liquid air
2. Natural gas
3. Iron
4. Around 450 °C
5. Around 200 atmospheres

EXAM QUESTION

a. To grow bigger/faster
b. Nitric acid
c. 35%

pages 28–29
The Manufacture of Sulfuric Acid

PROGRESS CHECK

1. Where nothing can enter or leave
2. When the rate of forward reaction = rate of reverse reaction
3. Vanadium(v) oxide
4. 450 °C
5. 2 atmospheres

EXAM QUESTION

a. Sulfur is burnt in oxygen
b. It is a reversible reaction
c. It increases the rate of reaction but has no affect on the position of equilibrium

Day 3

pages 30–31
Rates of Reaction

PROGRESS CHECK

1. They need to collide and when they do collide they must have enough energy to react (activation energy)
2. Increases the rate of reaction
3. Decreases the rate of reaction
4. Increases the rate of reaction
5. At the start

EXAM QUESTION

a. Catalysts
b. Increases
c. Decreases
d. Reactants

pages 32–33
Energy

PROGRESS CHECK

1. Exothermic reaction
2. Endothermic
3. Energy
4. A reversible reaction
5. Endothermic

EXAM QUESTION

a. Hydrated copper sulfate ⇌ anhydrous copper sulfate + water
b. Changes colour from blue to white
c. Add it to white anhydrous copper sulfate and it would turn blue

pages 34–35
Calorimetry

PROGRESS CHECK

1. Calorimetry
2. J
3. °C
4. g
5. 4.2 joules

EXAM QUESTION

a. Exothermic reaction
b. 84 J

pages 36–37
Energy Profile Diagrams

PROGRESS CHECK

1. Energy
2. The amount of energy to break the old bonds
3. Energy
4. Increase the rate of a reaction
5. They are not used up during a reaction

EXAM QUESTION

a. Energy profile diagram drawn with the reactants higher in energy than the products
b. The iron catalyst works by offering an alternative reaction pathway with lower activation energy

pages 38–39
Aluminium
PROGRESS CHECK
1. Aluminium oxide
2. Cryolite
3. No
4. Yes
5. Aluminium

EXAM QUESTION
a. $Al^{3+} + 3e^- \rightarrow Al$
b. $2O^{2-} \rightarrow O_2 + 4e^-$
c. The oxygen that is produced reacts to form carbon dioxide which erodes the electrode

pages 40–41
Sodium Chloride
PROGRESS CHECK
1. The production of fertilisers/as colouring agents/in fireworks
2. Barium sulfate
3. Barium chloride + sodium sulfate → barium sulfate + sodium chloride
4. Solid
5. Aqueous

EXAM QUESTION
a. Hydrogen
b. $2Cl^- \rightarrow Cl_2 + 2e^-$
c. Sterilise water, etc.

pages 42–43
Copper
PROGRESS CHECK
1. Heating
2. Electrolysis
3. Anode
4. Cathode
5. Cu^{2+}

EXAM QUESTION
a. Cu (s) → Cu2$^+$(aq) + 2e–
b. Cu^{2+}(aq) + $2e^-$ → Cu (s)

Day 4

pages 44–45
Acids and Bases
PROGRESS CHECK
1. A H^+ ion
2. OH^-
3. H_3O^+
4. Hydrochloric acid, sulfuric acid and nitric acid
5. Ammonia

EXAM QUESTION
a. Ethanoic acid, citric acid and carbonic acid
b. A strong acid is completely ionised in water while a weak acid is only partially ionised in water

pages 46–47
Making Salts
PROGRESS CHECK
1. Neutral
2. Alkali
3. Acid
4. Magnesium chloride
5. Calcium chloride

EXAM QUESTION
a. Calcium carbonate + hydrochloric acid → calcium chloride + water + carbon dioxide
b. Any number under 7 but probably 1 or 2

pages 48–49
Metals
PROGRESS CHECK
1. It is the attraction between positive metal ions and negative delocalised electrons
2. Alloys
3. Iron
4. Nickel
5. It turns limewater cloudy

EXAM QUESTION
a. Steel
b. Brass

pages 50–51
Useful Metals
PROGRESS CHECK
1. It is too soft
2. Alloy
3. Copper and zinc
4. Drinks cans/bicycles/ aeroplanes, etc
5. Rutile

EXAM QUESTION
a. A layer of aluminium oxide forms that prevents any further reaction.
b. The car body will be lighter so the car will have a better fuel economy/the car body will corrode less so it may last for longer
c. An aluminium car body will be more expensive to produce

pages 52–53
Iron and Steel
PROGRESS CHECK
1. Iron(III) oxide
2. Gold
3. Blast furnace
4. Reduction
5. 96%

EXAM QUESTION
a. Wrought iron
b. Low carbon steel
c. Cast iron
d. Iron(III) oxide

pages 54–55
Water
PROGRESS CHECK
1. A rock that contains water
2. To remove suspended particles like clay
3. To reduce levels of micro-organisms to acceptable levels
4. It uses a lot of energy
5. Nitrates

EXAM QUESTION
a. White precipitate
b. Silver chloride
c. Solid

Day 5

pages 56–57
Hard and Soft Water
PROGRESS CHECK
1. The Sun warms the sea
2. Carbon dioxide
3. Increases solubility

4. Decreases solubility
5. Increases solubility

EXAM QUESTION

Increasing the temperature decreases the amount of carbon dioxide that is dissolved in the drink

pages 58–59

Detergents

PROGRESS CHECK

1. To remove coloured stains
2. To remove stains at low temperatures
3. To make it more attractive
4. It helps the water to drain away
5. The maximum water temperature to use

EXAM QUESTION

a. To remove the dirt from the fabric
b. An acid
c. Hydrophilic end

pages 60–61

Special Materials

PROGRESS CHECK

1. Pencil lead/lubricant/electrodes
2. C_{60}
3. Fullerene, diamond and graphite
4. Lustrous and colourless
5. Black

EXAM QUESTION

a. Each carbon atom is bonded to four others by strong covalent bonds
b. Lots of strong covalent bonds
c. There are no free electrons/ions to move

pages 62–63

New Materials

PROGRESS CHECK

1. It is too big
2. Nylon
3. Teflon
4. To make non-stick saucepans
5. Bulletproof vests

EXAM QUESTION

a. It contains very small fibres
b. The small fibres trap air. This acts as a layer of insulation which stops body heat from being lost

pages 64–65

Vegetable Oils

PROGRESS CHECK

1. A and D
2. Sunflower oil and olive oil, etc.
3. Seeds, nuts and fruits
4. They have carbon double bonds
5. Vinegar and vegetable oil

EXAM QUESTION

a. Emulsion
b. Polyunsaturated
c. Biofuel
d. Bromine water

pages 66–67

Food Additives

PROGRESS CHECK

1. Throughout the European Union
2. Keeps unblendable liquids mixed together
3. Natural
4. Colouring
5. Antioxidant

EXAM QUESTION

1. Sodium hydrogen carbonate
2. To improve the texture
3. Carbon dioxide

pages 68–69

Crude Oil

PROGRESS CHECK

1. A compound that only contains carbon and hydrogen atoms
2. A part of crude oil/a group of molecules with a similar number of carbon atoms
3. At the top
4. Bitumen
5. Kerosene

EXAM QUESTION

a. Fractional distillation
b. Evaporates
c. Condense
d. Fractions

Day 6

pages 70–71

Fuels

PROGRESS CHECK

1. Hydrogen and carbon
2. Carbon dioxide and water vapour
3. Carbon
4. Global warming/greenhouse effect
5. It produces soot/carbon monoxide

EXAM QUESTION

1. Carbon dioxide
2. Carbon/soot/carbon monoxide/carbon dioxide
3. Water (vapour)

pages 72–73

Air and Air Pollution

PROGRESS CHECK

1. Nitrogen
2. Carbon dioxide
3. Plants evolved
4. Oxygen
5. Sulfur dioxide

EXAM QUESTION

a. Mars/Venus
b. Plants evolved
c. Global warming

pages 74–75

Pollution

PROGRESS CHECK

1. New jobs
2. Noise/dust/lorries
3. Plastic
4. It can produce toxic gases
5. It is hard to separate them out

EXAM QUESTION

a. It can absorb water
b. It is difficult for bacteria/mould

to grow so food stays fresh for longer

pages 76–77
Alkanes and Alkenes

PROGRESS CHECK
1. C_nH_{2n+2}
2. Fuels
3. (Covalent) bonds
4. Cracking
5. C_2H_4

EXAM QUESTION
a. Alkenes
b. It has carbon double bonds
c. Ethene + steam → ethanol

pages 78–79
Alcohols, Acids and Esters

PROGRESS CHECK
1. Alcohols
2. Esters
3. Alcohols
4. Carboxylic acids
5. Any number under 7

EXAM QUESTION
a. 7
b. Solvent/fuel/drinks etc.

Day 7

pages 80–81
Polymers

PROGRESS CHECK
1. Monomers
2. Polyethene (polythene)
3. Polypropene
4. It makes it more flexible
5. It will not soften but may eventually burn

EXAM QUESTION
1. Ethene
2. Alkenes
3. C_2H_4

pages 82–83
Limestone

PROGRESS CHECK
1. Calcium carbonate
2. Sedimentary
3. Calcium oxide (quicklime) and carbon dioxide
4. Quicklime
5. Slaked lime

EXAM QUESTION
1. Igneous
2. Marble

pages 84–85
Structure of the Earth

PROGRESS CHECK
1. Solid
2. Liquid
3. Mantle
4. Dozen/twelve
5. Tsunami

EXAM QUESTION
a. The crust and upper mantle
b. A few centimetres per year
c. Mountains formed as the Earth's crust shrank as it cooled down

pages 86–87
Cosmetics

PROGRESS CHECK
1. A mixture made when a solvent dissolves a solute
2. Nail varnish, etc
3. Ethanol
4. Jasmine/lavender, etc
5. Reacting an alcohol with a carboxylic acid

EXAM QUESTION
a. Fermentation
b. Glucose → ethanol + carbon dioxide

pages 88–89
Chemicals

PROGRESS CHECK
1. Hydrogen
2. Oxidising chemical
3. Toxic
4. Attacks and destroys living tissue
5. Orange/yellow

EXAM QUESTION
a. Carbon dioxide
b. Oxygen
c. Chlorine
d. Ammonia

Notes

Notes